# BLAMELESS
# CONTINUOUS
# INTEGRATION

# BLAMELESS
# CONTINUOUS
# INTEGRATION

*A Small Step Towards Psychological
Safety of Agile Teams*

## VIVEK GANESAN

PARTRIDGE

**To order additional copies of this book, contact**
Partridge India
000 800 10062 62
orders.india@partridgepublishing.com

www.partridgepublishing.com/india

# CONTENTS

Introduction ........................................................................ vii

Acknowledgments .............................................................. xv

Chapter 1 Practical Continuous Integration ........................ 1

Chapter 2 The Stimulus-Response Dance ........................... 16

Chapter 3 Effects Of Blame .............................................. 30

Chapter 4 Do Developers Control Build Results? ............... 39

Chapter 5 Handling Build Failures Without Blame ............. 53

Chapter 6 Building A Blameless Culture ............................ 63

Chapter 7 Sensible Information Radiators For Faster

           Response Times .............................................. 72

Chapter 8 Metrics And Course Correction .......................... 96

Appendix-A: 21 Instances Where A Wrong Person Gets

           Blamed For A Build Failure ........................... 108

Appendix-B: Low Cost Build Monitor Using Raspberry PI ... 114

Appendix-C: Tackling The Blame Game Dragon ................. 121

# INTRODUCTION

*English teacher* – This was the title of my first ever job. Though the job title commanded that I was supposed to be *teaching*, I, being my sloppy self, ended up *learning* something instead, most of the time on the job.

The reality has not changed much even today. I still continue to do things that are quite the opposite of my duties mentioned in my job description. Today, my job role is to be an Agile Change Agent for some incredibly bright teams in a cool startup. However, instead of changing the teams, I ended up changing *myself* and my outlook on certain aspects of life on the job. Oh, What a criminal I am!

Blameless Continuous Integration is one such idea that ended up changing me and my outlook on software development forever. This book is aimed to help agile teams and organizations in their journey of building impactful software without compromising on the *people* factor.

## Blameless Continuous Integration – What is it?

In the earlier waterfall days of software development, we used to have a certain time and budget allocated to a phase called 'integration'. This is the phase where the code produced by different teams used to be merged in order to create a coherently working software. The integration activity sounds easy unless one gets into the actual integration phase. Due to communication gaps and inconsistencies in design, the integration phase typically results in the discovery of numerous bugs termed as 'Integration issues'.

Since integration used to be towards the end of the lifecycle and near the release date, integration issues cause a lot of psychological stress to everyone involved in the system. Also, integration can sometimes run into several months because of the number of unknowns involved in the process. It is practically impossible to estimate how much time an integration activity will take before starting it. Due to all these reasons, this phase is fondly nicknamed as 'Integration Hell' by the software development community.

The Agile revolution had different plans for the integration heavy software. Continuous Integration or CI, in short, is a technical practice advocated by Extreme Programming (XP) philosophy as a way to build high-quality products in an iterative manner. XP says "If something is hard, do it often." Since integration is hard and is risky, in CI, the changes are integrated frequently to create a software build at least once in a day.

Wikipedia defines Continuous Integration (CI) as 'the practice of merging all developer working copies to a shared mainline several times a day.'

Continuous Integration typically involves a build server integrating changes according to a chosen frequency (or even each commit) and trying to create an integrated build. When the integrated code is built successfully, all is well. When the integrated code fails to build, the build is considered 'red' and the team is notified of the same to take corrective action.

The red build is a sign that the code is currently unusable. With time, came the advocacy for running local or private builds before submitting any change in the spirit of 'preventing' red builds. Even after such advocacy, the builds kept failing often and, in some organizations, almost all the time.

This gave rise to a culture of blaming the build breaker and sometimes, penalizing the build breaker when builds fail. This book is aimed at explaining why blaming the build breaker is not only a non-solution to the problem of failing builds, but also is plainly wrong. This book builds up a case of doing CI without blame, hence the name Blameless Continuous Integration.

The first few chapters explain how blaming the build breaker is ineffective and also invalid. In the later chapters, this book will also tell you practical suggestions that you can implement in your workplace right away to build a healthy culture of fixing builds as soon as possible, to minimize the trouble caused by the failed builds.

## Who is this book for?

This book aims to help four kinds of audience, at different levels.

**Developers & Testers:** If you are a developer or a tester in an agile team, you can use this book to understand some of the practicalities and common misconceptions while developing high-quality software at scale. This book starts with an introduction to Continuous Integration and builds upon the topic with practical variations of it. You will also learn what you can do to influence your team to stay away from blame games, in the context of agile software development.

**Change Agents:** If you are a Scrum Master, Agile Coach or any other variant of change agent, you can use this book to understand both tactical and strategic ways to influence the teams to practice Continuous Integration without blame games, negative emotions or stress. With this book, you can meaningfully guide the teams so that they traverse through the maturity curve towards Continuous Deployment, in the pursuit of reducing the time between concept and cash.

**Managers:** If you are a manager who serves development teams, this book will get you started with the thought process of how you can improve the morale of your developers by consciously avoiding blame in the context of Continuous Integration. This book will also help you by giving some practical advice on designing effective workplaces for agile teams and the metrics to track in order to evaluate the effectiveness of Continuous Integration practice in your organization.

**Leadership:** If you are a leader in an agile or yet-to-be agile organization and interested in technology too, this book will help you in getting a view of risk avoidance and customer focus from the perspective of the development teams. This book will help you in understanding the need to consciously avoid blame and the practical ways the teams can avoid integration risks early in the lifecycle by establishing effective workplace design and some meaningful practices.

## How to read this book?

This book contains eight chapters that are meant to be read sequentially. This book is written in a conversational tone and has a lot of questions directed at you, in between the paragraphs. Take time to answer those questions to yourselves to get an optimum experience of reading this book.

Additionally, there are a few questionnaires available in dedicated pages in a couple of chapters. I suggest you fill in these questionnaires when you encounter them before proceeding further. These questionnaires are designed to make you think on your own to arrive at conclusions before actually getting knee-deep into some of the concepts.

There is an appendix towards the end of this book. You can find independent articles and practical guides related to the subject matter there.

## How is this book organized?

This book has a total of eight chapters. Each chapter revolves around a central concept and when the chapters are read sequentially, they tell a coherent story.

The first chapter, *Practical Continuous Integration,* takes you through multiple flavors of CI followed by many organizations and ends with three acid tests for CI, in the context of Blameless CI.

The second chapter, *The Stimulus-Response Dance,* introduces Social Psychology and explains the different variations of stimuli that a CI system gives to the teams and how different people respond to those stimuli differently. It takes you through the actions, feelings and expressions that different people display when they encounter a failed build.

The third chapter, *Effects of Blame,* gets into details of how the blame affects the developers and the organization as a whole. This chapter also explains how blaming the developer for a failed build results in a behavior that is quite the opposite of what CI recommends, ultimately making the CI practice ineffective.

The fourth chapter, *Do Developers Control the Build Results?,* asks a pertinent question of whether a developer can completely control the build results, using an analogy of driving a car. This chapter also explains how a developer has no complete control over whether this code change will fail the build or not thereby proving that blaming the build breaker is not only ineffective but also invalid, in the first place.

The fifth chapter, *Handling Build Failures Without Blame*, explores the case where the developer's commit is the one that caused a particular build failure and possibilities of working without blame in this context. This chapter also creates a mental framework for classifying build failures into two categories and suggests effective action items that can be implemented when you encounter each category of build failure. This chapter also makes a case for built-in resilience in the CI process.

The sixth chapter, *Building a Blameless Culture*, explains what happens when developers don't fix builds. This chapter shares practical tips about how to build the 'fix immediately' culture while retaining the 'no blame' mindset by effective communication, behavior modeling and catalysis.

The seventh chapter, *Sensible Information Radiators for Faster Response Times*, takes a systematic view of the CI system and tries to list out ways in which one can effectively communicate the build failure to the developers. This chapter goes into the nitty-gritty of workplace design, in the context of information radiators. At the end of this chapter, you will have a collection of experiments that you can try in your workplace to influence your developers to fix builds fast.

The final chapter, *Metrics and Course Correction*, starts with listing out the measures that one should not measure in a Blameless CI workplace and proceeds to list the measures that make the most sense and how to use those measures to feed the continuous improvement process of the organization as a whole.

This book also has an appendix with independent write-ups that can add to the core concepts and practices mentioned in the earlier chapters.

Buckle your seat-belts and prepare for a journey that, I promise, will change at least one of your perceptions about the art of software development.

# ACKNOWLEDGMENTS

Anything ever created by a human being is not solely due to the efforts of its creator alone. Numerous beings, human or otherwise, contribute to the creation directly or indirectly. Likewise, the book 'Blameless Continuous Integration' was made possible by the efforts and support of numerous beings, whom I find pride to be associated with.

The most precious are those who help you in your times of hardship. I thank the efforts of my dearest friends Virag Dhulia, Jyoti Tiwari, Bhuvnesh Sharma and Rukma Chary, who have brought me out of despair in times of darkness and taught me the art of being peaceful in the face of adversity. Their encouragement made me a better and positive human being. This book would not have been possible if I had not met them in my life.

I would not have been able to even go to school if my dad did not work very hard. He is my default support system. He taught me the value of hard work. My mother has always taught me how to give the love back to the world. These values of hard work and giving back to the world are the top two motivators for me to document one of the things that I learned the hard way.

My brother, ten years younger to me, has taught me multiple lessons about changing one's own mindset, knowingly or unknowingly as I saw him growing from being a playful child to a responsible adult now. He finds comfort in situations of chaos and this got me thinking what he would do when he comes to my chaotic workplace. This hypothetical question generated a lot of ideas in me that I experimented with and have presented in this book.

A good work environment is one which believes in your strengths and gives you opportunities for experimentation of new ideas. Gainsight is simply the best place that I have ever worked so far. 'Success for all' is one of the values of Gainsight. My bosses Ashok Duguputi, Shrikant Kashid, Somasekhar Bobba and Sreedhar Peddineni allowed me to be original, which resulted in me trying a lot of experiments, which led me to this book. I am privileged to have worked with extremely talented and humble servant leaders and Scrum practitioners at Gainsight namely Gayathri Naik, Sowmiya Chellamuthu, Raja Kuditipudi, Vinay Kumar Konduri, Suresh Gunasekaran, Srinivas Muthyam, Kavitha Muneeswaran, Pawel Kumar Dalal, Bhavesh Kumar, Shashi Kant and the awesome Scrum teams of Gainsight. These individuals and teams asked questions that contributed to at least one sentence in this book. The development managers and architects at Gainsight, including but not limited to Swatantra Agrawal, Pradeep Saraswathi, Naveen Ithappu, Ankit Jain, Krishna Sekhar Sarangam, Murali Krishna, Varun Menon and Ahmed Jamal Maaz extended their co-operation to the experiments and discussions at work, without which I would not have learned much.

Like all software products have bugs, all books have errors in them. I am lucky to have a huge list of friends namely Swatantra Agrawal, Jyoti Dandona, Subba Reddy, Ranjith Tharayil, Lalit Das, Suresh Gunasekaran, Prashanth Kamasamudram, Ankit Jain, Pradip Ghosh and Bhuvnesh Sharma, who agreed to review this book's manuscript for errors. Believe me, this book would have looked nasty if these people did not help me.

Being the short-term action oriented guy that I am, the act of writing a book without any feedback for a very long time is not very motivating. This required me to appoint my besties Lalit Das and Suresh Gunasekaran as my 'progress bosses'. These people did well to remind me of my milestone due dates, sometimes going to the extent of pestering me to get into action. Without their help, I would probably have not finished writing even half of this book by next ten years. Pohar Baruah from Partridge Publishers also helped me in keeping track of time with his periodic follow-ups.

I thank my Scrum trainer Pete Deemer, who introduced this Big Data developer to the world of agility. I would not have started thinking about Blameless Continuous Integration if not for the agile conferences that I had the privilege to attend. I am indebted to the organizers and speakers of Agile India conferences, XP India Conferences, Discuss Agile conferences and Business Agility Roadshows for providing me the 'networking' opportunities and showering me with awareness about practices in multiple organizations.

I thank my *gurus* Rajshekhar Chittoory and Sanjay Kumar for teaching me certain aspects of systems thinking and coaching that went into the experiments, which formed the basis of this book.

All my acknowledgments will be useless if I do not acknowledge the contributions of Partridge Publishers. Their professionalism showed in every bit of interaction that I had with them. Their frequent follow ups kept me on track. They also deserve the praise for bringing this book to the world earlier than it would have been brought otherwise.

# Chapter – 1

# PRACTICAL CONTINUOUS INTEGRATION

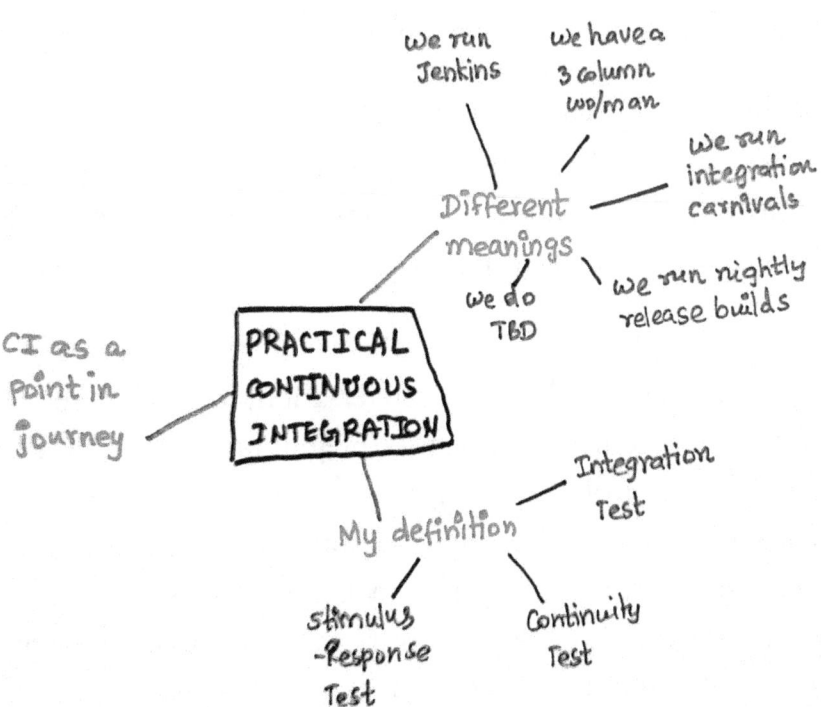

We run Jenkins

We have a 3 column wo/man

We run integration carnivals

We run nightly release builds

Different meanings

We do TBD

CI as a Point in journey

PRACTICAL CONTINUOUS INTEGRATION

Integration Test

My definition

Continuity Test

stimulus -Response Test

## Continuous Integration – Many Meanings

Have you been to agile conferences? If not, you must attend some of them. If you have attended some already, what is your favorite thing about these conferences?

To me, the networking breaks are one of the best things about agile conferences. One gets face-time with people belonging to different organizations and gets to see the world of work from the perspective of a different organization than the one s/he currently works with.

While I was experimenting with 'Blamelessness' as a philosophy, I felt that 'Continuous Integration' was one practice that needed the help of this philosophy the most.

After interacting with a wide variety of people belonging to organizations ranging from startups to enterprises, I was surprised to know that the term 'Continuous Integration' gets many different practical meanings depending on the organization's context.

In the middle of a hallway conversation, when I asked 'Do your teams practice Continuous Integration?', the answer that I heard, in most of the cases, is 'Yes, our teams do!'. However, when I asked some follow-up questions, I came across multiple flavors of Continuous Integration being followed in different organizations. This made me classify the Continuous Integration (CI) practitioners into five categories.

With some generalizations(but no judgments), the CI practitioners fall into five categories depending on how they answer the follow-up questions.

## Category – 1: We run Jenkins

This category of people answer "We run Jenkins" when you ask them "Can you tell me more about your CI practices?". Most importantly, all the sentences that follow "We run Jenkins" are about Jenkins plugins or about build monitors using Jenkins. They do not talk anything about their branching strategy or their release frequency.

This category of practitioners believe the 'tool = practice' school of thought. To them, Jenkins is CI and CI is Jenkins. They would run Jenkins, Travis or any other CI tool marketed in the *DevOps* arena and automatically assume that they practice the practice just because they use the tool.

If asked explicitly about their branching strategy, some of these people tell that each team works in a different branch and Jenkins validates every commit to those team branches. They also tell that they merge all the branches typically a week or two before the release date. This means that they use Jenkins just as a 'Commit Verification Machine' and not as an integrator.

This category is unlikely to reap benefits like early dependency resolutions, regression avoidance, etc because their integration (which they call merging) happens only once towards the end of the release.

This is similar to claiming that you are a cook just by buying an oven. In order to be a cook, you need to do more than just buying the oven, or even running it. You need to produce a consumable product – a food, in this case.

Similarly, in order to be a practitioner of CI, you need do more than just running a Jenkins server, which we will discuss at the end of this introduction.

## Category – 2: We have a 'Three Column Wo/Man'

This category of people answer "Of course, we have an integrator." when you ask "Can you tell me more about your CI practices?". When you ask them "Oh. What is an Integrator?", they would respond, "Integrator is a role that we assign to a person, typically a star developer. His job is to integrate all development branches to the release candidate branch."

Some organizations even seem to have an unofficial nickname to this role – 'The Three Column Wo/Man'. This person is often seen wandering in the team area with his IDE open, showing three columns (aka the diff merge window) and frequently asking individual developers to choose the lesser evil between the three/two options. Some not-so-social integrators can do this job using Slack channels or emails by passing around screenshots of the three columns.

In most of the cases, the integrator is not a full-time job. This means that the person has some other deliverables along with integration. In some cases, I hear that this integrator is the one who runs the unit tests before merging.

This structure is good for teams that are very small. This also uncovers key issues like dependency mismatches, regressions, etc. However, the moment the team size to integrator ratio goes high, we can visibly see some difficulties.

This integrator is prone to think of other developers as undisciplined because developers don't care about making the integrator's job easy by fixing test failures early on. Why do developers not care? Because, it is not their job. Their job is only to churn out code – a lot of code.

In some cases, the frequency of the merge is left to the discretion of the integrator with a single mandate – 'We need the merged code before the release'. Some integrators perform the integration frequently and some do it whenever their other job (Remember they are developers too?) permits it.

There is another uncomfortable truth – Only the integrator knows how to merge the code and no one else – of course, not always!

## Category – 3: We run Integration Carnivals

This category of people answer "All of our teams merge our code at the end of each sprint." when you ask "Can you tell me more about your CI practices?" When you ask them, "Who exactly does the merge?" they respond with "We rotate 'Mergers' in each team. They get together on the sprint closure day to merge their team's code to release branch."

With the increase in the number of teams(or branches to be specific) contributing to a product, this activity of 'merge' takes longer. This becomes unpredictable when there are merge conflicts that need to be resolved with the consultation of many teams, with each incoming merge.

Because of the amount of time these rituals consume and the buzz they generate, people tend to nickname them as *Integration Carnivals*. Some organizations try to reduce the time consumed by this activity by maintaining a single branch per program (a bunch of logically related teams).

Each team in the program checks-in to the same branch – acting like a mini trunk-based-development for the program. Each team would have a continuous commit-check job that validates every incoming commit.

In this setup, the management structure also follows the code merge dynamics (or is it vice-versa?) and this arrangement, more often than not, ends up in finger pointing between teams and/ or programs, in case of any issues during the carnival. The management tries to create a checklist of prerequisites for any branch to qualify for the merge – the one that includes code coverage benchmarks, feature summary, etc.

This flavor is very good for structured co-ordination between teams operating under different managers. This also provides the intended benefits of regression avoidance, early dependency checking, etc. albeit at a slow pace.

However, if this carnival is not taken seriously by the leadership and communicated as a useful activity to the teams, then the teams would treat this activity as a diversion. Code coverage becomes an afterthought. Teams would write tests on the last day of the sprint when the 'Merger' finds that their branch does not satisfy the checklist. We all know enough about the quality of tests written due to pressure.

## Category – 4: We do Nightly Release Builds

This category of people answer "Every team's code is merged automatically every night by our Jenkins to create a release candidate." When you ask them "What if you could not create a build due to merge conflicts or test failures?" they respond saying

"We try to fix the failures and resolve conflicts first thing in the morning to get a build out as soon as possible."

This structure achieves the biggest benefit of CI – fixing dependency issues and regressions early – closer in time to when they were introduced, when compared with the previous categories. Within a short time of introducing daily merges, the teams typically express an intent to write more tests, naturally increasing the coverage. Since improvements are out there to see, typically, the end to end automated tests also get more attention (and investment) as the time progresses.

Since merge is done after a vert short interval(a day), the merge conflicts resolution would not take a lot of time, even if merge conflicts occur. Since people are 'in context' with the code that they wrote the previous day, the build failures can be fixed in shorter time.

However, doing this is not as easy as it sounds. This needs a clear commitment from management and the teams that daily builds are THE most important things for any team. In some organizational cultures, the teams need to have a rotation mechanism for 'build monitor and fixer' who monitors the last night's build and fixes in the morning. The same person cannot do this job every time as this might involve arriving at the office earlier than usual.

If there is a politically charged workplace, this practice will make it even more charged with flame emails flying about which team broke the build the previous night. This blaming will have multiple side-effects, which we will discuss in a future chapter.

## Category – 5: We do Trunk Based Development

To be honest, I have never personally met anyone who does Trunk Based Development at scale, though I have read about some organizations doing it. But, I feel urged to introduce this category for the sake of completeness. So, based on the blogs that I have read about Trunk Based Development, if I ask these invisible people "Tell me more about your CI practices.", I assume that they would respond saying "We do Trunk based development."

Each member of the organization commits to the trunk – a single branch. Each commit is integrated and validated by Jenkins or any other CI system as soon as it is pushed. In order to not break things, developers typically write tests first. Regressions are identified immediately within minutes of pushing the commit.

However, doing this is not so easy. This requires a decently decoupled architecture and a culture open to vulnerability. Initially, the organization might end up not having any deployable builds for days. But, the situation slowly improves as people adapt to the new system. This causes a dip in the velocity (Oops! I uttered the taboo word!) of the teams as they have to learn to deliver at pace with the new system.

## What does this book mean by Continuous Integration?

If the industry recognizes at least five practical meanings of 'Continuous Integration', then which flavor is 'Blameless CI' about? In other words, when I say 'Blameless CI', which variety of Continuous Integration am I talking about?

Answering this question requires me to put my own interpretation of what Continuous Integration means. Rather than delving into the variations and then deciding which variations suit my perception, I would prefer to define Continuous Integration using three acid tests.

In the context of 'Blameless CI', (not sure if this is same elsewhere too), any team/organization that passes ALL of the following acid tests are considered to be as practicing Continuous Integration.

## Test 1 – Integration Test

To pass this test, you just need to answer 'Yes' to the following question:

Do multiple code-bases/branches get merged/integrated at some point in time?

a) Yes b) No

The integration could be at commit level or at branch level. But, there must be an activity (manual or automatic) to merge two different code bases and to check if the two different codes work well together.

## Test 2 – Continuity Test

Yes, I am just splitting the words in the term 'Continuous Integration' but this is something that I have seen some teams consistently missing to do before claiming 'We do CI'.

To pass the continuity test, the activity of integration should never have an end date, unless the product/project has an end date saying "We are not going to work on this product/project anymore." Also, the integration should be continuous with some cadence established – at every commit in some organizations, every night in some others, every week in some cases, etc. High frequencies lead to better results.

Anyone who passes this test will never have an intention of stopping the integration activity until the project/product's end of life.

### Test 3 – Stimulus-Response Test

This acid test talks about the 'people' aspect of Continuous Integration practice. When you do Continuous Integration, things go wrong – integrations fail due to compilation errors, merge conflicts, test case failures and a plethora of uncontrollable factors.

In order to pass this test, there should be a culture of responding quickly to the stimuli created by the CI system. This does not mean that there should be no 'red' integration builds but there should be a clear intention to fix 'red' builds as highest priority.

If you integrate every sprint but your teams take 'fixing builds' as a low priority, resulting in an integrated working code several days later, then you don't practice Continuous Integration, in my opinion. Why? It is simple to reason this out – You seem to value busy work more than addressing an openly visible release risk that stops you from getting an integrated working product.

## Continuous Integration – A Point in a Journey

Personally, I see Continuous Integration as a point (or a stop) in a journey of the train called organization. Traversing this stop of Continuous Integration is necessary to get to more exciting and valuable stops.

CI leads you to Continuous Delivery (CD), where you make sure that the product can be released reliably at any time and can be deployed with the push of a button. This would mean that you have all your unit tests, integration tests and automated end-end tests run on commits and a 'deployable' created successfully in very short cycles. Any errors in the CD pipeline will be treated as the highest priority interrupt and will be fixed faster than anything else.

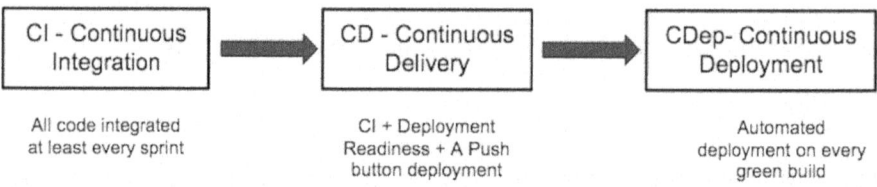

Further down the road, you can go to the Continuous Deployment (CDep) stop, for which, traversing the CD stop is necessary. Continuous Deployment eliminates the 'push button' in the CD. If all your changes are validated and deployed automatically, then you are in CDep.

Though I tend to think of this journey of CI, CD and CDep as having three stops, there are meaningful alternate schools of thought where CD and CDep are seen as one and the same. Also, some part of the agile community treats Continuous Delivery as a collection of practical aspects of one of the twelve agile principles – *Our highest priority is to satisfy the customer through early and continuous delivery of valuable software.*

## What is special about Blameless CI?

While the teams I served were practicing Continuous Integration at some level, I came across many situations where someone blaming the build breakers resulted in the behaviors that went against the very goals of Continuous Integration. For example, the developers started committing code less frequently in large chunks of changes, whereas CI advocates frequent and small commits.

This led to an experiment where we consciously avoided blaming the developer but still practiced CI. This experiment helped me open my eyes and question some assumptions that I had for years about software development at scale. By experimenting with different mental models, very late in my journey, I learnt that a developer never has full control over whether his commit can break the build or not. This made me understand that blaming a build breaker is not only negative but also invalid, in the first place.

Blameless Continuous Integration, the book that you are reading now is a compendium of the theories and practices that I learnt and sometimes, concocted while experimenting with teams practicing Continuous Integration. Here, the practice was intentionally done with a single guiding principle – not blaming the build breaker for any reason whatsoever.

The next chapter talks about what happens on the office floor when there is a CI culture and some behaviors which can undermine the very purpose of Continuous Integration practice in your organization.

**Summary**

While the definition of Continuous Integration (CI) is the same in theory, there are many practical flavors acknowledged by the industry. Each flavor has its own merits and challenges. Due to variations in practical meaning, we, for the context of this book, devised three acid tests to check if an organization is practicing CI namely, Integration test, Continuity test and Stimulus-Response test. Continuous Integration(CI) is also seen as the first stop in the journey towards Continuous Delivery(CD) and Continuous Deployment(CDep).

**Further Readings and References**

1. Continuous Integration: Improving Software Quality and Reducing Risk – https://www.amazon.com/Continuous-Integration-Improving-Software-Reducing/dp/0321336380

2. Jenkins: The Definitive Guide: Continuous Integration for the masses – https://www.amazon.com/Jenkins-Definitive-Continuous-Integration-Masses-ebook/dp/B005EI8686

3. Nightly Release Builds – http://softwareengineering.stack exchange.com/questions/56490/what-does-nightly-builds-mean

4. Trunk Based Development – A 5 minute overview – https://trunkbaseddevelopment.com/5-min-overview/

5. An apologist's defense of Trunk Based Development by tuesdayDeveloper blog – http://tuesdaydeveloper.com/2015/05/11/an-apologists-defense-of-trunk-based-development.html

6. Continuous build is not continuous integration by Dan North – https://dannorth.net/2006/03/22/continuous-build-is-not-continuous-integration/

7. Continuous Delivery: Reliable Software Releases Through Build, Test and Deployment Automation – https://www.amazon.com/dp/0321601912?tag=contindelive-20

8. Continuous Deployment – https://www.agilealliance.org/glossary/continuous-deployment/

9. A discussion in Stackoverflow forum about differences among CI, CD and CDep – http://stackoverflow.com/questions/28608015/continuous-integration-vs-continuous-delivery-vs-continuous-deployment/28628086#28628086

# Chapter – 2

# THE STIMULUS-RESPONSE DANCE

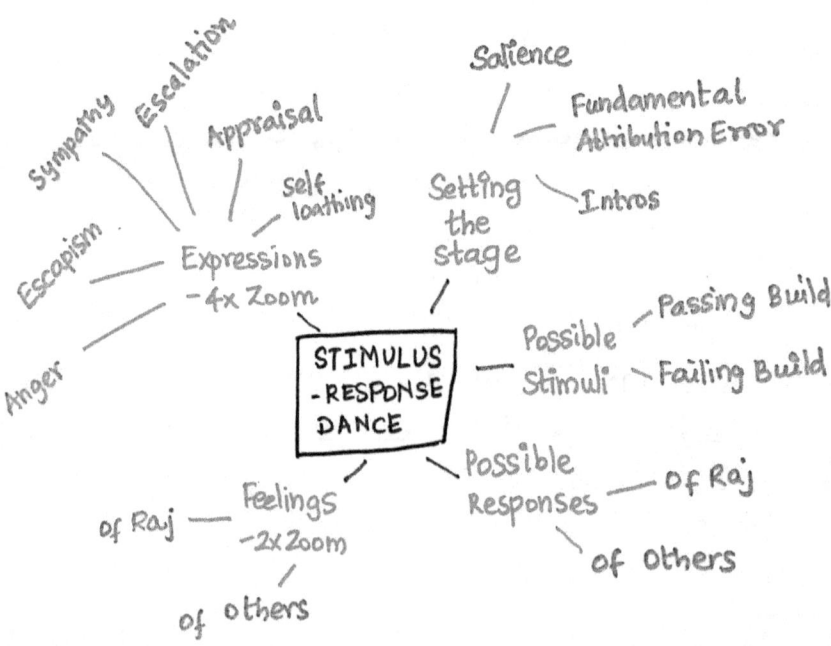

Psychology, as a field, is very vast. It involves different specializations such as clinical psychology, neuropsychology, social psychology, child psychology, etc. Among all these specializations, there is one specialization that we will use many times in this book – The Social Psychology.

**What is Social Psychology?**

The Wikipedia definition of 'Social Psychology' calls it as the scientific study of how people's thoughts, feelings, and behaviors are influenced by the actual, imagined, or implied presence of others.

Social Psychology is not only an interesting field but also a relevant field while dealing with software development teams. Software development teams are nothing but sets of people who work towards common shared goals(*hopefully!*).

This chapter is aimed at providing you with a basic understanding of some common errors of judgment that all humans commit, according to social psychology and then illustrate the case of an organization where you can see these errors manifest in practice. Let us start with the common errors of judgment that humans unknowingly commit.

**Social Salience and its effect**

Imagine that you work for an organization that develops and sells banking software. There is a very popular and helpful team called Team X in your organization. Everyday, when you enter your office, you need to pass through the work area of team X in order to reach your cubicle.

Two weeks ago, when you entered the office in the morning, you noticed a new relatively young face in the team X and you, being a not-so-socially-awkward person, proceed to greet the new person and introduce yourselves as Mr. or Ms. yada yada. The new person, in return, introduced himself as Mr. just-out-of-college and proceeded to explain that he would be contributing code to the team from the day one.

Today, you happen to hear in a water-cooler gossip that the team X had created unusually high number of bugs in the past iteration. What would be your thoughts about this situation? Care to record it below in a few words of your own?

---

---

---

---

Were you one of those people who thought about the new team member being the reason for sudden spike in the number of bugs? You might not be surprised to know that many people would have come to the same conclusion. If your answer is not the same, you might still find it valuable to understand why many people would have come to that conclusion.

Now let us analyze if we are correct in relating the high number of bugs to a new and novice programmer in the team X. Do you know anything about whether the team X committed to more work or less work than usual in the iteration in question? Do you know whether their product owner changed the scope of the story in the middle of iteration? Can any other situation lead to increased number of bugs? Yes? Then why did we (yes me too) relate the new developer to the increase in bug count?

Social psychology comes to our rescue by explaining that human beings tend to focus on 'salient' things when looking for explanations. The term 'salient' means 'something that draws attention'. The salience could be due to a variety of reasons – something standing right in front of you, something very different from the surroundings, something that just arrived, something that you were already looking for, etc.

In the case of team X, the salience of the new developer is because of the fact that he was 'new' and just-out-of-college in contrast to the other experienced developers in the team X. Since the new developer is a 'salient' feature of the team, you tend to associate the high bug count to the new developer. Your reasoning is that the team X did not have as many bugs when the new developer was not there. However, this may be right or not right depending on the additional information that you get.

Human beings are lazy(for a good reason) and they want to make judgments with as less information as possible. Hence, we choose to not ask follow-up questions about the other aspects that could lead to high bug count and instead directly judge the new developer as the culprit.

Hold on this concept of salience for sometime and we will recall this soon.

**Fundamental Attribution Error**

*The eldest man in a particular village developed a dislike for a particular farmer. This particular farmer knocked the old man's door every two days, requesting to mediate an issue between him and his neighbor.*

*Every time, the issue was the same. The neighbor's goat was grazing in the farmer's fields and the farmer accused the neighbor of being careless and demanded compensation.*

*One day, the neighbor knocked the old man's door. Surprisingly, this time, the neighbor, not the farmer, requested the mediation. Contrary to the usual norm, the issue this time was that the farmer's goat had eaten the vegetables from the neighbor's garden. The neighbor accused the farmer of being careless. Now, the farmer fell into his own trap.*

*The old man asked the farmer, "What do you say in your defense?". The farmer instantly replied, "What do I do, sir? My goat seems to have learnt from his goat!"*

This is not just a story but a common occurrence in our daily lives. The farmer called it carelessness when the neighbor's goat did something wrong and he called it 'situation' when his own goat did the same. This is what we call as Fundamental attribution error in Social Psychology.

When others do something wrong, we attribute it to their bad nature or intentions but when we do something wrong, we attribute it to the situation at hand. This is a common error of judgment that almost all human beings exhibit.

This becomes more pronounced when this fundamental attribution error uses the salience effect discussed earlier to find someone guilty.

Let us keep the theoretical social psychology for a while and proceed towards understanding how people behave in an organization that practices Continuous Integration.

## The Third Acid Test

Do you remember the third acid test (that we talked about in the last chapter) to validate if an organization really practices Continuous Integration or not? Yes! It is the Stimulus-Response Test? This third test validates whether people actually care about fixing builds as the high priority.

This chapter and the remainder of this book exclusively deal with the practical aspects of the Stimulus-Response dance that happens in an organization, in the context of Continuous Integration.

## Meet the Fictional Organization

In order to entertain you with an illustration, let me present you a fictional organization that has a practice of Continuous Integration (which passes all the three acid tests) in place.

For the sake of simplicity, let us assume that the fictional organization's teams have automated the process of integration by using Jenkins. This would mean that they have a Jenkins job that runs and passes when everything goes right i.e., when new code is integrated with the old code, compilation passes, unit tests run without any failures, blah, blah and blah. The integration job fails when something goes wrong. An instance of such a job is called as a 'build'.

Let us also assume that the organization has some communication mechanism to let people know whether the latest integration build passed or failed. I am purposefully giving less details about this communication mechanism because the details of the mechanism does not matter for this chapter. We have a future chapter that talks exclusively about how to effectively design this communication mechanism for optimum results. So, if you have any lingering questions about this organization's build status communication mechanism, hold on for some more time!

Oh! A bit of local jargon too – The organization calls all the 'passed' builds as 'Green builds' and all the failed builds as 'Red builds'.

## The Stimulus-Response Dance

Let us now lean back and watch an interesting social behavior similar to a complex group dance as we try to observe the Stimulus-Response combinations possible in this system with a microscope, at multiple zoom levels.

Let us take a first look at the system that we need to observe. At the center of this system that we analyze is a collection of complex organisms called human beings. We are interested in how the behavior of these organisms is influenced by the stimuli that they receive. Here, stimuli is nothing but the information that is passed from the system to the human beings.

What stimulus do the human beings get from the builds directly? In other words, if a build has just finished, what information do the humans get from the build? Can we categorize the possible variations in the stimulus and try to understand how the response of the human system varies accordingly? Let's go ahead and do just that!

## Meet Raj – The Developer

If you are into example based understanding, let me introduce you to the protagonist of our story today – Raj. Raj is a developer working for one of the many teams in our fictional organization. He joined the organization as a coder, just out of college, only a few months ago.

As of now, he committed a change to the code base and is waiting for the Jenkins integration job to finish. Raj is an important part of the system under observation now.

In order to understand the system under observation, let us list down the possible variations in stimuli and see how the system reacts accordingly.

## Variation 1 – Green Build

## Stimulus

The build is green. Raj's change integrated successfully with the current code in the branch.

**Response(s)**

i.  **Raj:** Raj is satisfied that his day's work is done and he goes on to do something else – his music lessons, perhaps!
ii. **Other actors:** Nobody else notices anything. Only player involved in this stimulus-response dance is just Raj.

This is a simple and straight-forward case.

## Variation 2 – Red Build

**Stimulus**

The build is red. Raj's change did not integrate successfully with the current code in the branch.

**Response(s)**

i.  **Raj:** Raj looks at the job logs and finds out what is wrong. He commits new code, aiming to fix the build failure.
ii. **Other actors:**
    a.  **Other developers:** The other developers hold back their commits waiting for the build to turn green. Else, they wouldn't know if their own commit is breaking the build or Raj's commit which broke the build just now. They are blocked until Raj fixes the build.
    b.  **Testers:** Testers (if the role separately exists) hold back testing the new code. No one wants to test the code that breaks the build. No one wants to test the outdated code too. They are blocked until Raj fixes the build.

## 2x Zoom – What about the feelings?

There are more actors than other developers and testers in the system and that is where the dance gets interesting.

I, purposefully, did not include 'feelings' in the above narrative of the response. I included only the 'actions'. If I include the feelings too, the responses would look more or less like the ones below. Let us do a 2x zoom on the responses.

### Response(s)

i. **Raj:** Raj feels terrible for having broken a build. He regrets having blocked many others. He comes to term with the fact and tries to fix the build as fast as he can, before someone points a finger at him.

ii. **Other actors:**
   a. **Other developers:** The other developers are irritated that the branch is red. They hate the fact that they need to wait (without any ETA) for Raj to fix the build in order to commit their code. They feel angry about Raj, who committed the code without taking care about whether it would break the build or not.
   b. **Testers:** Testers are uncomfortable with the latest development. They hate to get delayed on their testing just because Raj broke the build. They are concerned about the possibility that a junior developer like Raj could halt the entire team because of his carelessness.

Basically, everyone is furious. Wouldn't the manager notice when everyone is furious? Yes, he would! But, one thing at a time! We will talk about the manager's responses in a short while.

**4x Zoom – What about the expressions?**

Now, we are getting another level deeper into the responses. First, we talked about the 'actions'. Next, we went a level deeper and talked about the 'feelings'. Now, we will see how these feelings are expressed as 'expressions' by doing a 4x zoom.

We will organize this section a little different this time. Since expressions differ from person to person based on their personality, let us see some (limited) variations of how people would express themselves in this situation. Here is a list of some possible expressions.

1. Raj broke the build. What a careless guy! – *Anger*
2. Why should I fix it? It is Raj who broke the build! – *Escapism*
3. **From:** escalator@example.com
   **To:** manager_of_raj@example.com
   **Sub:** I AM BLOCKED
   Dear Boss,
   I am BLOCKED until Raj fixes the build. Why should I suffer for someone else's mistakes?

   Sincerely,
   The Escalator

   *- This is a formal escalation*

4. Raj, poor soul! He needs some more training. – *Sympathy*
5. Raj, you are a good developer. But, you could have done better at avoiding some build failures *this year*! – *Performance appraisal aka career nightmare*
6. I am an idiot! How could I have been this careless? – *Self-loathing*

There could be even more varieties of expressions that one can see in the teams practicing CI.

## Social salience of Raj

Looking at the responses for a red build, do you find a common idea that forms the basis of all of them? What is an implicit assumption that all of the responders seem to have made?

If you look closely, all the responders, including Raj himself, have assumed one thing – Breaking the build was a mistake, that too, on Raj's part. In other words, they attribute blame to Raj for the build failure and hold him accountable for the inconvenience.

Now, let me remind you the concept of 'social salience' again. Is it possible that everyone assumed Raj's mistake because of the salience that Raj's commit was the only difference between earlier 'green' build and the now 'red' build? Is there any other non-salient difference between the earlier and current builds? Did they take time to understand why the build failed? No!

Is salience clouding our judgment? It is quite possible that the salience has clouded everyone's judgment in the illustration.

## Is there a fundamental attribution error?

Now, in this example, the latest change was committed by Raj, a junior developer. What if it was an architect who committed it or a senior developer? In a culture of blame, there is a huge (but not certain) chance that the senior developer blames the situation for his failure – like the earlier tests were wrong, the database connection went down during the build, etc.

Do you remember anything when you see this behavior? Yes! This is the Fundamental attribution error that we talked about in the beginning of this chapter. The senior developer blames the latest committer when the build breaks after someone else's commit but blames the situation when the build breaks after his own commit.

This book's fourth chapter is dedicated to challenging the assumption that it was Raj's fault but let us not get ahead of ourselves. The next chapter is going to help you in understanding the outcomes of blaming Raj and how these outcomes affect the organization as a whole.

## Summary

Social Psychology warns us of judgmental errors that human beings are prone to. When build passes after a developer commits his code, typically no one notices. However, when a build fails, multiple people respond with blame towards the committer. Could this blame be due to the judgmental errors that Social Psychology warns us about?

## Further Readings and References

1. Social Psychology – Wikipedia https://en.wikipedia.org/wiki/Social_psychology
2. Effect of salience – https://en.wikipedia.org/wiki/Social_salience
3. Fundamental Attribution Error – https://en.wikipedia.org/wiki/Fundamental_attribution_error

# Chapter – 3

# EFFECTS OF BLAME

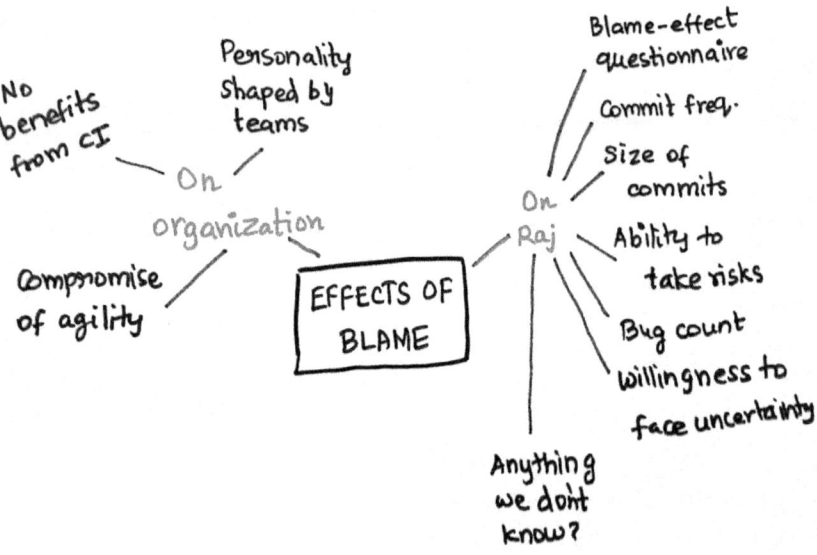

In the last chapter, we talked about the different types of stimulus-response patterns in case of red builds resulting in blame. Is this behavior of 'blame' reasonable? What effects does this blame produce? How does this blame affect the organization? This chapter aims to answer these questions.

## Effects of Blame on Raj

Without getting into the judgment of whether this blame is good or bad, let us try to see the effects of the blame on Raj. In the next page, is a questionnaire that I would urge you to fill, thinking about the situation at hand.

Try filling the questionnaire before reading further. This questionnaire is designed to make you think deeper than what we have already done, about the current situation.

## BLAME – EFFECT QUESTIONNAIRE

1. **Frequency of Commits:**
   If there is a culture of 'blame the breaker', Raj, the developer, would tend to push his code _____.
   a) frequently b) not so frequently

2. **Size of Commits:**
   Raj's commits tend to be _____.
   a) small b) huge

3. **Ability to take risks:**
   Raj will tend to _____ taking risks.
   a) enjoy b) avoid

4. **Bug Count:**
   Raj's code will have _____ bugs than a developer who has never ever broken a build so far.
   a) more b) fewer

5. **Willingness to face uncertainty:**
   Raj will be _____willing to work on an
   uncertain item than a developer who has never broken a
   build so far.
   a) more b) less

6. **Anything that we do not know about:**
   Can you think of any other effects that Raj could have
   based on his personality/upbringing/situation at hand?

   _____

   _____

   _____

   _____

   Have you written your answers? Did the questions make
   you think more about Raj as a person? Yes, that is one of
   the purposes of this questionnaire. Let us now talk about
   each of the questions and their implications to the work
   culture of the organization.

**Frequency of commits:** Remember? Nobody notices anything
when we have green builds. When a green build goes unnoticed
but a red build is frowned upon, the developers are pushed into
a special practice called FDD – Fear Driven Development.

From the developer's point of view, each commit is an opportunity
to integrate. This would also mean that each commit is an
opportunity to create a red or green build.

The developers are afraid to push changes because by pushing a
commit, they are opening up a chance for only ridicule and not

appreciation (because no one notices green builds). This results in infrequent commits. Developers with fierce managers commit at a frequency that depends upon when their managers go on vacations.

Hence, if there is a culture of 'blame the breaker', Raj, the developer, would tend to push his code <u>not so frequently</u>.

**Size of commits:** We just saw that Raj commits code infrequently. However, he has to write the same number of lines of code, with or without blame. Otherwise, he will not be able to deliver anything and he will easily come under everyone's radar for not being so productive.

This leads to a behavior that I call as 'commit batching'. Raj writes lot of code and commits infrequently. This leads to bulky commits instead of small ones.

Hence, Raj's commits tend to be <u>huge</u>.

**Ability to take risks:** Raj lives in an environment that does not tolerate failure. He does not have the necessary psychological safety to try out a different design pattern, a new framework or, if I take it to an extreme, even think on his own.

Raj tends to typically mimic what other senior developers do in the organization. He tends to copy-paste the code and designs because he can always point at 'existing' code when someone asks him about failures and say, "I thought my code would pass since another code with similar design is already there and passing."

Raj also tries to shy away from any new work that requires original thinking. He is happy with repeating the same design patterns for different user stories. If a requirement leads to untested approach, he tries not to participate.

In general, Raj will tend to <u>avoid</u> taking risks.

**Bug Count:** Remember? Raj batches his commits! This means that the commits don't go anywhere from his laptop for days.

Late commits result in the code being untested for a long time. When the code is not tested immediately, this results in lots of bugs. Another truth is that it takes longer to fix those bugs because Raj is not 'in context' when the bugs are discovered. He would have moved on to his next work and he has to do context switching (a costly mental activity) to fix the bugs from the code that he wrote long back.

If you take the case of a developer who has never had a build failure, his confidence levels are likely to be high about his code. He does not suffer from the 'Infrequent commits' disorder because he does not expect a reprimand due to a red build. Since this other developer commits frequently, testing also happens immediately. This results in less bugs.

This is why Raj's code will have <u>more</u> bugs than a developer who has never ever broken a build so far.

**Willingness to face uncertainty:** What will Raj do when his team's product owner gives him just the high level acceptance criteria for stories? Raj will push for even finer criteria (otherwise known as test cases) that talks about very obvious scenarios too because he does not want to be wrong about any assumption that could break a build.

Raj does not trust his own intuition and will ask the product owner to write down acceptance criteria involving even the simplest validation scenarios. Doing this is not wrong at all! However, preferring 'written' documents over conversations and using them later to prove one's innocence only shows lack of trust.

Did I tell you that Raj wants these low level tests (I am not going to call them acceptance criteria anymore!) written *before* the sprint planning? Raj will not commit to anything that does not have 'certainty' of going well. This typically extends the sprint planning and grooming sessions, wasting precious time of team members.

Raj also is less willing to volunteer to work on research stories or spikes because they lack the clarity. Anything that can open up a chance of a failure is a strict no-no for Raj.

Thus, Raj will be <u>less</u> willing to work on an uncertain item than a developer who has never ever broken a build so far.

**Anything that we do not know about:** Remember the description I gave about Raj when I introduced him? No? Here it is: "Raj is a developer working for one of the many teams in our fictional organization. He joined the organization as a coder, just out of college, only a few months ago."

We know very less about Raj so far. Even with such less knowledge, we were able to bring out the ill effects on Raj when he is blamed for build failures. What do we not know?

- Do we know whether Raj is going through relationship issues?
- Do we know whether Raj is suffering from health issues?
- Do we know whether Raj has ailing parents?
- Do we know whether Raj earns enough money to pay off his student loans?
- Do we know if Raj works part-time somewhere else to make ends meet?

We need to realize that no amount of information is enough to decide what effect this toxic blaming environment has on Raj.

**How does this affect the organization?**

If we understand that each team in the organization is made up of people like Raj, then the effect of this 'build breaker blaming' on the whole organization can be understood clearly.

Each team's personality is shaped by its team members and interactions between them. What do you think of a team that commits code infrequently, creates more bugs and is not ready for any risks or uncertainty?

Similarly, the organization's personality is shaped by its teams and the interactions between teams. What do you think of an organization that creates lots of bugs and does not take any risks?

Isn't the practice of continuous integration expected to reduce bugs and risks by integrating frequently? If blaming Raj takes these two key benefits of continuous integration away from the organization, why do we even need to practice continuous integration?

We need to understand one thing – the entire organization's agility is compromised by encouraging the blame on the person who broke the build.

## If blaming is bad, what's the alternative?

When you show a solution for a known problem to a man, he has a solution but when you show an unknown problem without a solution, you have created a problem for him. This book does not want to create problems for you.

This book tries to attempt at giving practical solutions to the problem of 'Build breaker blaming'. The solutions that we arrive at are based on some common principles rooted in social psychology and organizational behavior.

In the next chapter, you will learn about the amount of control a developer has on whether his commit will break the integration build or not and why this information matters to any organization doing CI. Ready to challenge your own perceptions? Move ahead to the next chapter.

## Summary

Blaming the build breaker creates behaviors contradictory to CI like low frequency commits, bulky changes, less willingness to take risks, high number of bugs, etc.

## Further Readings and References

1. The No Asshole rule – https://en.wikipedia.org/wiki/The No_Asshole_Rule
2. Why I added a "Peson (sic) to Blame" field to our bug tracking system – https://news.ycombinator.com/item?id=4179298
3. Software Engineering StackExchange discussion on the same "Person to blame" field – http://softwareengineering.stackexchange.com/questions/154733/my-boss-decided-to-add-a-person-to-blame-field-to-every-bug-report-how-can-i
4. Rituals of Shaming in the Software Industry – https://dev.to/pavsaund/rituals-of-shaming-in-the-software-industry

# Chapter – 4

# DO DEVELOPERS CONTROL BUILD RESULTS?

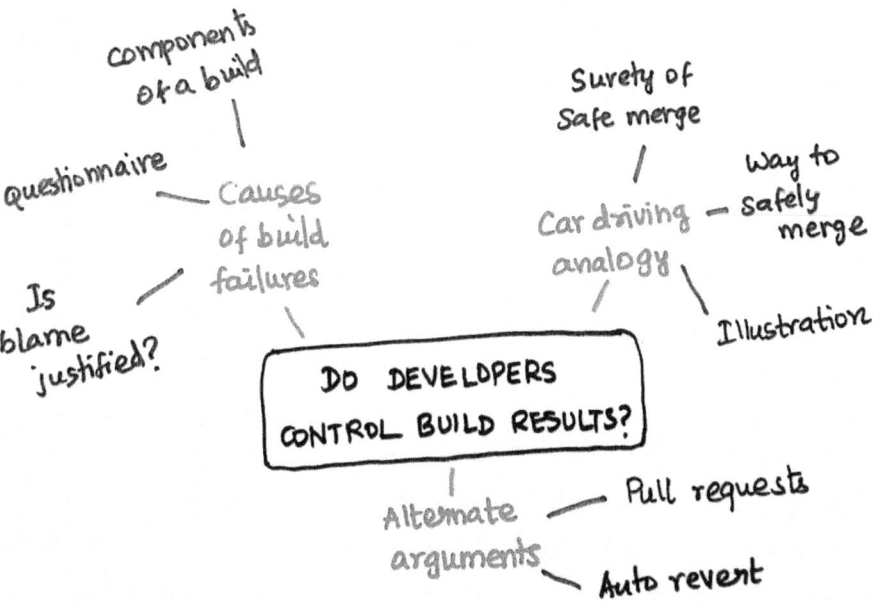

## The Road Intersection

Do you know how to drive a car? Even if you do not know, you must have at least travelled by a car. Imagine that you are driving your car from home. Your home is in a small town and you are traveling to the nearby city, where you will picnic for the day. Given, in the next figure, is a map of one particularly dangerous point on your way to the next city.

The road from your home is narrow and it needs to merge into a highway that connects your road to the city. The highway is a very busy one. There is only one small problem – exactly at the merge, a tall wall completely blocks the visibility of the highway from your road. Similarly, any vehicle, like the massive giant truck you see in the image, on the highway cannot see that your car is about to drive into the merge.

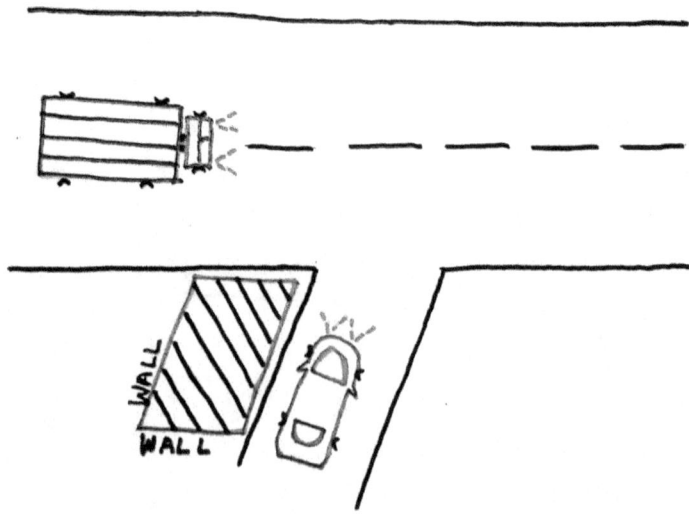

Now that you are about to drive from your road onto the highway, try to answer this question:

Can you be 100 percent sure about driving into the highway without causing an accident?

a) Yes   b) No

There is a high chance that you answered "No". Assuming that you are a non-ninja human driver, you would certainly have answered 'No'. I have a good news for you: You just admitted that developer does not have any control over whether his commit will break the build or not.

## Integration Build Analogy

I lied a little (only a little, I promise) when I was narrating the situation about you going to picnic. You were not driving a car but instead you were going to push your commit. The car symbolized your commit.

The narrow road from your home was nothing but the local branch in your laptop. The broad highway was the remote branch that many other team members, along with you, push their commits to.

The massive truck was the commit that someone else pushed to the remote branch, just after you checked-out the latest code from the remote branch.

The road merge was nothing but the integration of your code with the already existing code in the remote branch, especially with the truck.

Remember the wall that blocked the visibility? It was nothing but your ignorance (or is it innocence?) about a truck being committed by your teammate after you checked out. Yes, it was the wall of invisibility.

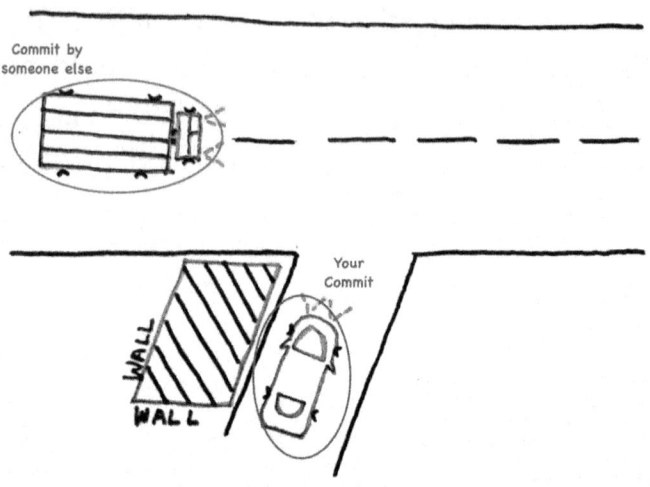

Now, the last piece of analogy – the question you just answered. The question that I wanted to ask you was actually this:

Can you be 100 percent sure about integrating your code with the code in the remote branch(this contains extra commits that came in after you checked out) without causing a build failure?

a) Yes   b) No

Did you just answer 'No' to this question earlier? Good news – feel comforted that your answer is not correct, this time and the earlier time too.

## How to get 100% control?

Now, to those of you who answered 'Yes' to the car question, you were right! If merging into a highway without visibility was impossible, such road merges wouldn't exist in the real world. But, we see such merges a lot of times in the real world.

How do the 'Yes' drivers make sure that they avoid the accident all the time? They seem to know a trick that the 'No' people do not know. Let us hear that secret trick from the 'Yes' people.

The 'Yes' people slow down or even stop their car just before the merge, take a peek into the highway, proceed into the merge if no other vehicle is approaching. If they see another vehicle approaching, they wait until it passes and then proceed driving their car towards the merge. Isn't this obvious?

## How to avoid a build failure?

Now, let us translate the 'Yes' driver trick to the code merge analogy. Like a good driver, before pushing your commit, you need to check if anyone else committed after you checked out. If no one has pushed their commits, you can commit your code. If someone has committed, pull the new commit, integrate with your code in your local machine and then push the code. *Tada*!

**Note:** I am assuming that you run tests before every push, that way, the code you push is a green code.

*Vivek Ganesan*

**Are you still hundred percent sure?**

I understand that the 'yes' people are sure about avoiding commits. But, are they hundred percent sure? What if a this happens – before driving into merge, I stop and find that no vehicle is coming on highway but by the time I drive into the merge, a Ferrari comes on the road at its top speed?

The 'yes' people made an important assumption: No vehicle will come on the road between the time they check the highway and the time they drive into the highway. If the assumption is broken, their chances of causing an accident are same as 'no' people. Here, the 'yes' people need to slow down to the point of immobility if they are particular about safety.

Similarly, in the code merge scenario, we made an assumption that no commit will be pushed by anyone else while we locally integrate the just-now-identified new commit on the remote branch with our yet-to-be-committed code. What if your teammates push code at a high rate that by the time your local tests finish running, your code is stale again because another person has already pushed his code? Here, you need to slow down to the point of immobility if you are particular about not breaking a build. I am sure nobody wants their developers to slow down to the point of immobility.

The time of engineers is more valuable. Here we are talking about multiple engineers. It is more costly to expect a green build all the time than to let the builds fail and then fix immediately. If you are still not convinced, let me use a practical illustration to explain this fact.

## Illustration

Let us consider that the CI build takes 10 minutes to complete. If we are expecting the developer not to break the build anytime, he needs to run the equivalent of a CI build locally, which will also take 10 minutes.

Let us say that Raj has to do a tiny change that takes almost no time. He checks out the latest code, makes the change (in no time) and starts running the local build in his laptop. He cannot check-in the code for another 10 minutes since that is the time the local build takes to complete.

Is there a possibility that his current build will become irrelevant in any case? Yes there is a possibility! If another developer pushes a change while he is running the build, Raj's ongoing local build is irrelevant because he is running the build on an outdated version of the code. In case to comply with the 'always green' rule, Raj will have to stop his build (assuming he knows that a push happened from another developer), pull the latest code thereby integrating with his local change, start the local build again and pray to The Almighty that no one pushes another change in the meantime. If the release date is nearing, there is a high possibility that his prayer will go unanswered.

In short, if the rate of pushing commits is high enough to continuously make the local builds irrelevant, Raj ends up in an indefinite catch-up game. Not only Raj, the entire clan of developers end up in this catch-up game.

Towards the end of this book, there is an appendix which lists down 21 instances (non-exhaustive) where a wrong person can be blamed for a build failure. Make sure you take a look at it.

**Why not try a Pull Request culture?**

With the emergence of Github as a preferred hosting and source control system for many open source projects, a new way of working also emerged – The Pull Requests. What are these pull requests? Authors work on their own copies called forks and raise a pull request to the maintainer once they are done. Instead of the author merging the changes to the remote branch, the maintainer of the project merges the author's code to the main branch, after reviewing the code.

In addition to pre-commit review, the Github ecosystem also provides support to verify the commits in the pull request using different CI tools. This flow is aimed at one thing that we are calling as impossible – The Evergreen Build. If this workflow promises an evergreen build, why not switch to this pull request workflow?

Pull request workflow has a very visible bottleneck – the reviewer/maintainer. This workflow was designed keeping open-source projects in mind. In an open source project, people contribute at their own pace, during their own free time. Velocity is typically not a focus. Rate of incoming changes is typically low when compared with enterprise products, which are developed in dedicated development centers with salaried contributors.

In case of enterprise product development, a lot of people are committing changes at any given time. There is a high velocity of incoming changes. So, even when you go to pull request workflow, due to high rate of changes, you will still have the same cycle time because you have just transferred the job of doing a local build from a developer to a CI server that can work with Github pull requests.

**What else can cause build failures?**

If you try to divide at the CI build system into smaller components, you will most probably end up at a component list similar to the one below.

1.  Build machine
2.  The code base before integrating the current change
3.  The current change – Yet to be integrated
4.  Tests (Both in current change and already existing ones)
5.  Test data
6.  Build/Test environment – Third party services, Databases, etc

Your list may be slightly different from mine but I want you to think about how many of the above items can Raj completely control when he is the author of the current change, yet to be integrated? Record your observations in the following questionnaire in the next page. For each item, take time to think whether Raj can predict/control that item to influence the result of the integration build.

# QUESTIONNAIRE TO ASSESS DEVELOPER'S CONTROL ON BUILD COMPONENTS

For each of the following factors, select an option that you agree with the most, assuming Raj is the author of the current change, yet to be integrated.

1. Raj has _____ direct control over the build machine, the hardware/virtual machine on which the integration build runs.
   a) No   b) Partial   c) Full

2. Raj has _____ direct control over the state of the code in the remote branch before integrating his code, including the commits just pushed by someone else.
   a) No   b) Partial   c) Full

3. Raj has _____ direct control over his current change, the one yet to be integrated.
   a) No   b) Partial   c) Full

4. Raj has _____ direct control over the tests, both his new ones and the ones that already exist before his change in the code base.
   a) No   b) Partial   c) Full

5. Raj has _____ direct control over the test data used by both his new tests and the ones that already exist.
   a) No   b) Partial   c) Full

6. Raj has _____ direct control over the third party services like databases used by the integration build.
   a) No b) Partial c) Full

Have you finished filling your answers in the questionnaire? All the people who filled this questionnaire sitting next to me gave an answer of 'full' to only one question above – the one about Raj's own current change. All other answers were either 'no' control or 'partial' control. Yes, Raj has full control on only one out of at least six components of a build.

Until this point, we were assuming that only the integration issues, that too particularly, the incompatibility of current code with the earlier code, result in build failures. But, when you look at the CI build system as a total of its components, a build can fail because of wrong state of any of these components. How many times have we seen a build fail because the test data was wrong or the third party service was unreachable?

In summary, a build can fail because of six things out of which only one is in the control of the developer but we still blame the developer for any build failure. Isn't this just grossly wrong?

One thing that we all need to understand is this: Unless we dig into details, we do not know why a build failed. Committer is just one of the six causes for build failures. Blaming the committer is due to the 'salience' (remember the time we talked about *social salience*?) of the last commit where the last commit could have been perfectly running fine before merge.

## The case of auto-revert

What stops us from auto-reverting the latest commit when a build fails? This will make sure that we will always have a green build. Great idea, right? *No!*

Anyone who favors auto-revert has not understood the point behind filling the questionnaire. We just saw that many things, not only the commit, lead to build failure. When we favor the auto-revert solution, it only means that we are still in the belief system that commit is the only thing that can break a build.

Typically, the auto-revert also creates a bigger problem when the build environment has caused a build to fail. In such case, the system will revert and commit and since the environment is to blame, the revert commit will also fail the CI. This will prompt the system to revert the revert, then again revert the revert that reverted the revert and so on, doing busy work, consuming power and processor cycles but playing a practical joke on the audience.

I hear that Chromium's development team is into reverting the failing commits. I am curious how they do it without any shortcomings. There must be a solution! What is the fun in me being right? But until I get a satisfactory answer to the Chromium question, I stand by my advice not to try auto-reverts.

## What if the commit's author wants to revert?

Yes! That is a good news. This means that the author understands that his commit is the reason for failure and needs some more time to work on the commit. More importantly, he does not want to block others and hence takes this decision consciously. No one else reverts the author's commit. You have got a developer who realizes that fixing a build is important, that too in a right way without taking shortcuts.

In short, unless the author of the change wants to revert for his/her own reasons, any other person reverting the code would just equate to blame and assumption about the author being wrong.

**What about the third item?**

We have been concentrating on the cases where something other than the developer's code caused a build failure. Intentionally! With tongue in cheek! I figured that at least some of you will see through my fraud of not talking about developer's commit breaking the build, till now! They say that there is a time and space for everything. The time and space for talking about developer's commit breaking the build is now and the next chapter, in the same order.

**Summary**

Developers do not have full control over the build results. Blaming the developer for a build failure is not just bad but also invalid. In order to ensure a green build all the time, a developer needs to slow down to the point of immobility. Pull request cultures do not support high commit velocity environments and reverting the commit is also not an option since it implies that the commit caused the failure, which may not be the case.

**Further Readings and References**

1. Blameless Postmortems and a Just Culture – https://code ascraft.com/2012/05/22/blameless-postmortems/
2. A lighter pull request workflow – https://hackernoon.com/ a-lighter-pull-request-workflow-972301e30c5

# Chapter – 5

# HANDLING BUILD FAILURES WITHOUT BLAME

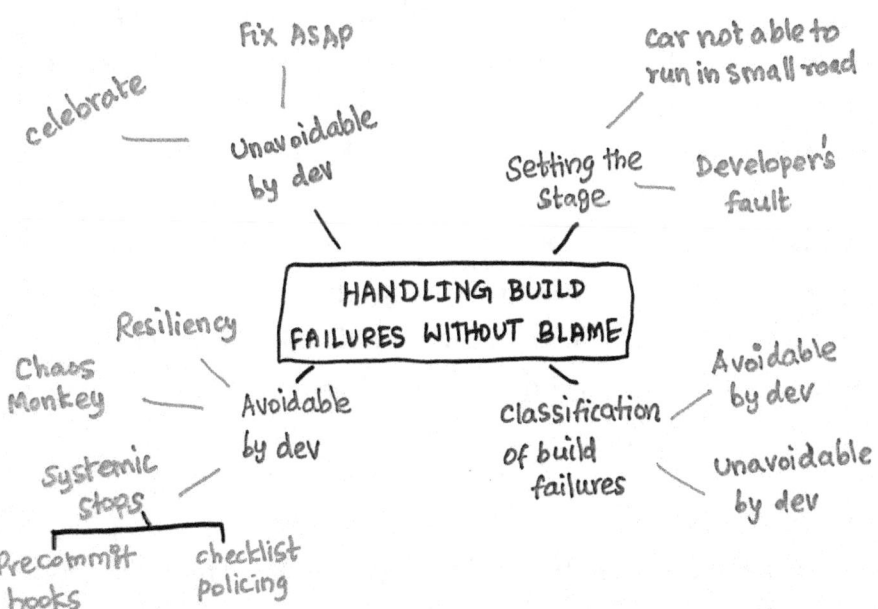

## A ridiculous piece of joke

In the last chapter, I rambled on and on about how committing code is like driving into a road merge without visibility. Poor Raj driving a small car from his house cannot get into the highway without slowing down to the point of immobility. Poor Raj! Some of you might be thinking that this must be a ridiculous piece of joke in the form of a well-meaning book. Some of you might even think that this book is a well-planned conspiracy by build-breaking inept developers to excuse themselves from the embarrassment.

Yes! Go ahead and ask your question. Yes, please. Ah! I know your question: What if the car couldn't run properly even before the merge? What if you were totally drunk and were trying to get the car onto the highway through the merge? Why do I not talk anything about the possibility of the driver being drunk before the build broke? Oops sorry! Before the car crashed?

Such a pity! I have been rambling on and on in the entire last chapter about how it is not the developer fault that the build failed. I even went ahead and made a list of components of build system. Here is a reproduction of that sneaky list in case you want to see it again.

1. Build machine
2. The code base before integrating the current change
3. The current change – Yet to be integrated
4. Tests (Both in current change and already existing ones)
5. Test data
6. Build/Test environment – Third party services, Databases, etc

While I wrote in length about how the above items are not in Raj's control, except for item number 3 – his own change yet to be integrated.

What if Raj's code could not compile even in his laptop? What if he conveniently forgot to type a semicolon in the Java file and compilation fails in the CI server after the merge? Should we not blame Raj for that incident?

No!

You heard it right – No! We shouldn't blame Raj for that. Read on to know why this answer is not ridiculous.

## Classification of build failures

Let us try to use a simple system to classify build failures.

Typically, the common mental framework of people forces them to classify build failures into two buckets:

1. Avoidable
2. Unavoidable

This classification is incomplete because it does not qualify who can avoid the build failure. If we change this classification slightly to reflect who can avoid the build failure, we will end up with the following buckets.

1. Avoidable by developer
2. Unavoidable by developer

A missing semicolon is avoidable by developer. However, an integration issue or failure of a third party service like a database is not. Effectively, if the cause of build failure is the developer's commit by itself we can call it as avoidable by developer. Otherwise, we will call it as unavoidable by the developer.

This classification provides you a mental framework to arrive at the action items while dealing with them.

## How to deal with a failure, avoidable by Raj?

Believe me or not! A developer's commit solely breaking a build is a critical feedback to the whole system. For that matter, any single individual having a possibility of causing a negative stress on the system itself is a critical feedback.

## The Chaos Monkey Phenomenon

A number of reputed tech companies, which have businesses worth millions and billions of dollars developed a sudden interest in a software called Chaos Monkey. Suddenly, Chaos Monkey was the cool software sought after by the 'cloud ninjas' of the world.

What does the Chaos Monkey do? It just does what its name says. It creates chaos in the production environments of its users. If you install Chaos Monkey in your organization, it would randomly take some of your production servers down. Yes, *production servers!*

Why do people voluntarily deploy it when it is known to cause problems to working software? They do it because it helps them find loopholes in their resiliency systems. The acid test for their resiliency is whether the application stands the test of the chaos monkey's stealthily planned outage.

The user group of this Chaos Monkey is of the opinion that no one server should have the power of causing havoc to the total system by turning itself off. When the app fails because of a single server's outage by Chaos Monkey, the engineering teams take it as a personal insult and make sure that the dependency is taken care of.

Why am I talking about Chaos Monkey here? Just to divert you from your point that I am avoiding to talk about Raj's missing semicolon? No!

If your organization is very serious about its Continuous Integration practice, then a chaos monkey like Raj should not be able to take a ride on it. No no! I am not asking you to fire Raj. Nor I am asking you to check the semicolons in code tests of all the candidates you interview. I am just asking the organization to be resilient.

## Being Resilient

We must remember one thing: *Things happen because the system allows them to happen.* Raj is able to check-in code that doesn't even pass the basic compilation because the system allows him to do that.

When we say that the system should not allow avoidable faulty commits, you can approach it in two ways.

1. Reprimand Raj and give him a warning not to commit avoidably faulty code next time.
2. Make it systemically impossible for *any* developer, not only Raj, to commit an avoidably faulty code.

## Approach-1: Reprimand Raj

The first approach of reprimanding Raj is the most common approach. This is the gut-feel approach that people implement because it is easy. However, this is not very effective. This approach makes Raj a loser and pushes him back into the 'blamed' Raj state. Remember once we discussed about how blaming Raj spoils the whole purpose of Continuous Integration? Yes? Then, don't blame Raj for this too. Period.

Even if Raj takes the blame sportively and accepts his mistake, this approach is not likely to prevent anyone other than Raj from doing this mistake. As long as the other developers think that only Raj did a mistake and not get themselves into the situation of reprimand, people would not realize their carelessness. This would mean that potentially every developer needs to be given the reprimand for this approach to take effect in all developers. Oh! Did I tell you that you might be hiring new people in between? The job of the manager will quickly turn into that of reprimanding one engineer every single day.

At the end of all this, developers get defensive because they are afraid now, commit code less frequently and in bulky chunks of changes. Back to the square one! Congratulations!

## Approach-2: Systemic Stop

If you want to systemically stop the avoidably faulty commits from getting in, you could use the help of either technology or the process. I personally prefer the technology route and only like to use the process route where technology is infeasible.

Technology route is called Pre-commit hooks and the process route is called checklist policing.

**Pre-commit hooks:** If you are in the Git world, you need to look at pre-commit hooks if you have not done that yet. Even if you are not in the Git world, you could find ways to mimic a pre-commit hook. Pre-commit hooks are nothing but scripts that will run before every commit. This would mean that the commit will not happen when the hook fails.

You could store a pre-commit hook in your code base itself and configure the pre-commit hook to run just the compilation before committing.

This way you are giving a safety-net for all the developers. Even if they make an avoidable error, it wouldn't affect others. It will fail in their own local computer, saving all the embarrassment.

But, beware of a caveat. When you try to misuse pre-commit hooks to suit your obsession for green builds by making the pre-commit hook run the entire CI build locally, developers will surely find a way to disable it. They would not disable it because it takes more time but because it takes more time doing an 'irrelevant' build most of the times – Remember? Long local builds get irrelevant in days of high velocity commits.

What if now every developer has stopped missing semi-colon but keeps making some other avoidable mistake very often? Simple! Add a check for that common mistake to the pre-commit hook! However, my personal thumb rule is that when pre-commit hook takes more than a few minutes, the developer gets impatient and hacks around it anyways! It is important to understand that the pre-commit hook is to help the developer and not to delay the developer.

**Checklist policing:** This is the art that drives many beautiful things in agility. This art is found in everything ranging from Definition of Done to Acceptance Criteria. In this context, checklist policing means that you need to create a checklist of things to check before committing and ask the developer to fill it and if you don't trust the developer much, ask his change to be cross-checked by another team member (like peer-review) before pushing every commit.

This method also can be suited to drive continuous improvement just like the pre-commit hooks by adding items to the checklist or removing items from the checklist depending on the need. Similarly, this should not be used to bloat the process of committing code. Remember KISS! Keep it simple silly!

**The Bottom Line**

There is one message that I deeply care about: *Blame does not solve anything.* No developer comes to office everyday planning to break a build. No one gets pleasure out of build failures.

If we deeply care about our workplaces and the people in our teams, we need to know how to use the system to prevent avoidable mistakes, in code or otherwise.

## How to deal with failure unavoidable by developer?

This is the case where we do not find a punching bag. If the build failure is due to an integration issue, which is nothing but incompatibility of one developer's code with another's, what can we do now?

## Build Failure – A Reason to Celebrate

In case the build failed due to incompatibility of one developer's code with another's, Voila! You have caught an integration issue early. Isn't this the whole point of doing Continuous Integration?

We can look at this in another perspective too – that of communication. An integration issue is typically a communication gap between two developers or two teams. An integrations issue signals that two entities that must have been communicating to each other were not doing so. It could also mean that the level of information communicated between the teams is not sufficient.

Hence, it is justified to have only one emotion when you see a build fail due to incompatibility or integration issues – joy. You can only celebrate the fact that the practice of CI is giving its fruits. But, don't let the celebration take long because you need to fix the build as well. I would even go to the extent of fixing the build first and then celebrating later. Would you also please?

## Summary

Build failure can be either avoidable by developer or unavoidable by developer. Avoidable failures arise because the system allows them. The avoidable failures must be stopped from within the system (Pre-commit hooks, checklists, etc). Unavoidable failures are reasons to celebrate, but only after fixing the build.

## Further Readings and References

1. Chaos Monkey – https://github.com/Netflix/chaosmonkey
2. A sample pre-commit hook – https://github.com/git/git/blob/master/templates/hooks--pre-commit.sample

# Chapter – 6

# BUILDING A BLAMELESS CULTURE

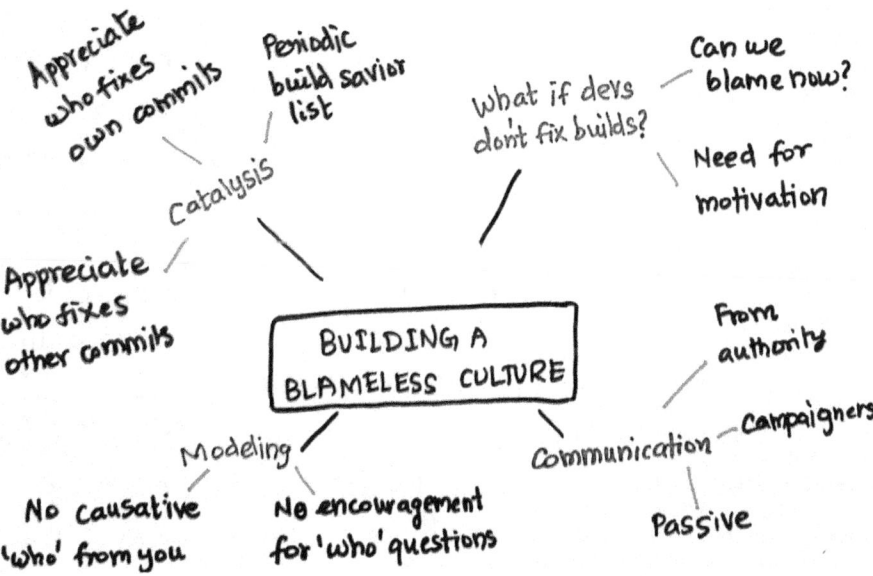

Appreciate who fixes own commits

Periodic build savior list

Catalysis

What if devs don't fix builds?

Can we blame now?

Need for motivation

Appreciate who fixes other commits

BUILDING A BLAMELESS CULTURE

From authority

Modeling

Communication — Campaigners

No causative 'who' from you

No encouragement for 'who' questions

Passive

## New Way of Work

Now, after reading almost half of this book, you have realized that blaming developers for build failures is wrong. You even have convinced your peers and higher ups that they should also not indulge in build-breaker-shaming. All of them have agreed to your line of thought even when they initially called it as bullshit, after you showed them how your argument was actually correct. Everyone takes an oath in the company's biggest conference room that no one will shame a build breaker from the next day.

Here comes the next day. All developers are happy because no one would blame them. Morale is at an all time high! Smiling faces everywhere and the floor is electrified. You are proud to have started such a noble movement.

At around noon, when people are engrossed in their work, a build fails. You come to know of it from an email that was triggered from Jenkins. Managers catch themselves having the intention to find out who broke the build but they hold themselves back. They, being true to their oath, choose to wait until the build is fixed.

Fast forward 4 hours, it is 4 PM now and no one seems to have fixed the build. You get an email from a frustrated tester saying, "It is not working! Should we go back to old ways?" Oh yes! All your peers and your manager are copied in that email. You sit in your cubicle with shooting blood pressure, massaging your forehead and wanting to go and hide in the bathroom. You are hoping for someone to come and save you from this gigantic embarrassment.

## Can we blame the developer now?

Now what should you do? Blame the developer for not fixing the build? Yes, technically, you are not violating any of the principles discussed so far by blaming the developer. You are not blaming him for breaking the build. You are blaming him for not fixing the build ASAP. But, I am sure you will understand by now that blaming the developer for whatever reason will not change the system and can only lead to degeneration of the system as a whole.

Thus, you are alone in a situation where you have nobody to blame but yourself because your organization is stuck due to your decision of buying a stupid book, reading half-way through it and talking to your peers about something that sounded bullshit right from the word 'go'.

Now that I have made you aware that you will end up as a joke if you follow the advice of Blameless CI without much scrutiny, let me also take the opportunity to introduce you to the tricks of motivating the developers to fix builds. The rest of this chapter is going to be a collection of practical ways in which you can motivate the developers to fix builds. If you are the developer yourself, you can still motivate other developers using these methods.

**Methods to motivate developers to fix builds**

The methods that one can use to bring in a change in mindset can be classified into three categories.

1. Communication
2. Modeling
3. Catalysis

## Category 1 – Communication

In many situations, when you end up not getting results on the first day of the change, the problem is mostly that the target audience did not even know of the change. It is essential to communicate the intention of moving to Blameless CI to the entire organization. Without doing this, any effort to practice Blameless CI will end up in a disaster. Doing Blameless CI without proper communication is similar to expecting a woman to decide the place of your wedding even before you have proposed your love to her.

Let us list out some of the effective ways of communicating this change. You could do a mix and match of any of the methods described here to communicate your intention to change.

1. **Mass communication from authority:** An email or a blog post detailing the change from an authority, say CTO of your company, carries a lot of attention and weightage. This could even be a shiny notice signed by your CTO on your office notice board. This sends an invisible message that the company is serious about this change.

2. **Campaigners:** You can also pick a campaign group of people who can convince others about the Blameless CI change and can talk about the intended results. Let them wear a cool looking t-shirt or a badge reading "Ask me about Blameless CI" for a day or two. This way, anyone can approach a campaigner and strike a conversation to understand the change being talked about. In order to step up the involvement, you can even make the campaign group to act like a brotherhood, that gives the cool t-shirt

to anyone who demonstrates the spirit of Blameless CI flawlessly.

3. **Passive Communication:** Passive communication happens when no one is pushing the communication but people learn the news when they want to learn it, asynchronously. This method takes some time to build up a movement but it is intrinsically motivated hence effective too. This can be done by pasting non-intrusive posters with one liners that create curiosity about the change, all over the team work areas. You could even add a note at the end to visit a particular web page in company intranet to get more details about the initiative.

## Category 2 – Modeling

It is not enough to just let the news about the change float and still do the same old thing anyways. It is essential for you to model the expected behavior to let the people know that Blameless CI is not the same old wine in a new bottle.

Here are some ways to model the behavior so that the teams can be inspired by you.

1. **No 'who' question from you while looking for cause:** Stop yourself from asking the 'who' question whenever a build breaks. Never ask anyone who broke the build. Believe me, anyone who understands Blameless CI will not use the words 'Broke the build' as an accusation. If you ask me who broke the build, I would always give you the same answer every single time: The System. A build failure is a feedback to the system. That is what it is. So,

don't ask the 'who' question. When you ask the 'who' question, others will think it is okay to ask too.

We are not stopping ourselves from blaming because of our benevolence. We are stopping ourselves from blaming because it makes *business sense*. So, don't make it sound like 'I understand. People mess up. But that is okay'.

It is also important that this rule applies only when you are looking for the 'cause' of a problem. However, it is good to ask 'who will fix the build'. Here, you are not looking for the cause but the result hence encouraging action rather than promoting blame.

2.  **No encouragement for 'who' questions while looking for cause:** Now that you have stopped asking 'who' questions while looking for cause of failure, you should stop encouraging others while they do it. You cannot just get into an argument with people about why they are wrong to point others out. You can just humbly ask them, "Will it be useful to find out who is ready to fix this instead?" with a sincerely smiling face. Yes! Sincerely. Smiling. Face. No hard feelings. Accept that the mindset change takes conscious effort and is not achievable with just a single communication.

## Category 3 – Catalysis

After you have communicated the intention to change and have started modeling the behavior to inspire others, it is essential to create an environment favorable for the targeted culture to be nurtured. This is where catalysis comes into picture. In this phase, you end up being a catalyst for the mindset change and the practice of Blameless CI.

Below are some methods in which you can try to create an environment conducive to Blameless CI.

1. **Appreciate if anyone fixes failures after other's commits:** This is to remove the mind block of people that they need to fix builds only when their commit is involved.

   If a developer commits code, gets an emergency call from his family and has to leave, without waiting for the CI build to complete, we need someone to fix the build when he is gone. We do not want to call him while he is attending a family emergency. Appreciate the person who fixes failures after other's commits, with even the minimum capacity that you have. Even a handwritten thank you note from someone influential or authoritative will be a great motivator. This approach will create a positive momentum when appreciation is instant and heartfelt without any sugar coating.

2. **Appreciate if anyone fixes failures after their own commits**: Appreciate a person who fixes failures even if they happen to be after his own commits. Now, this is just a logical extension of Blameless CI. There is only one situation where you would *not* appreciate this person – when you believe that he himself was a culprit. If you

understand that the developer does not have control over the build result, it is only logical to appreciate anyone who fixes the build irrespective of who committed before the build failure.

Remember! Your focus is to create an environment where people will strive to fix the failing builds as soon as possible.

3. **Periodic Build Saviors List:** Compile a list of all people who fixed the builds in a given period – a week, a sprint or a month and ask someone up in the chain send a public or organization-wide email sincerely thanking these people. For those who feel that methods (1) and (2) above appear to be over-enthusiastic and too much heavy, you could just do this one thing to catalyze the mindset change.

Getting appreciation from someone in top management can give a positive kick to the developers. You don't necessarily involve monetary benefits into this because involving money can actually make the people trick the system or at the worst, get demotivated as money is the worst extrinsic motivator for work. Human Intrinsic Motivation is a broad topic which is beyond the scope of this book. If you are curious, I would strongly recommend you to pick up and read the book 'Drive' by Dan Pink.

**This is not all!**

Are the above the only methods to communicate, model and catalyze? Is the communicate-model-catalyze the only model with this you can approach this mindset change? Not really! You can always come up with your own methods and models that will work in your organization's context. Yes, you can use your own creativity and skills to create a solution. The list in this chapter is not an exhaustive list.

Take some time and write down five things that you could think of in order to promote the Blameless CI mindset in your organization.

1. _____

2. _____

3. _____

4. _____

5. _____

## Summary

Build-fix culture does not come in a day. You cannot blame the developer for not fixing the build either. You can improve the developer motivation to fix the builds by using communication, behavior modeling and catalysis.

## Further Readings and References

1. Communicating to drive culture change – http://knowledge.senndelaney.com/docs/thought_papers/pdf/communications_2014.pdf

2. Drive: The Surprising Truth about What Motivates Us – https://www.amazon.com/Drive-Surprising-Truth-About-Motivates-ebook/dp/B0033TI4BW

# Chapter – 7

# SENSIBLE INFORMATION RADIATORS FOR FASTER RESPONSE TIMES

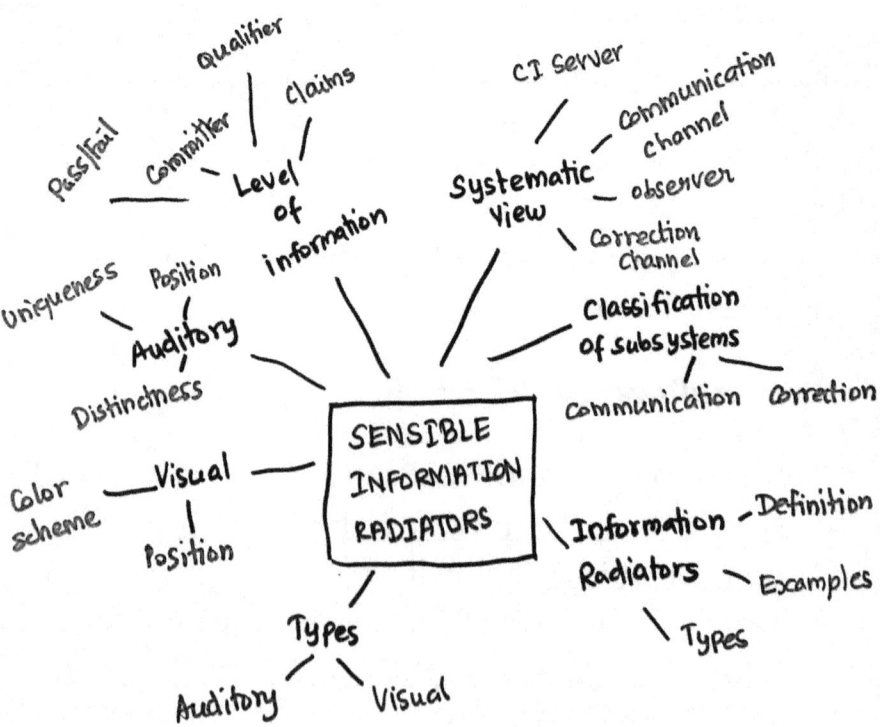

## Response time – The prime focus

You must have understood from the previous chapters that fixing the build failures as quick as possible is very important for Continuous Integration to work. This chapter aims at providing practical advice on how to improve the response time of a developer, which is the prime focus in keeping a CI system operational.

## A Systematic View

Systems thinking is a method that is used to generate creative solutions to problems by looking at problems as a result of interaction between different parts of a system. We will adopt systems thinking view to arrive at ways in which we can optimize the response time to build failures.

If I have to draw the entire CI setup in a minimalistic diagram, I would draw it the following way.

The CI system can be considered to be made of the following subsystems.

1. The Build server
2. The communication channel
3. Observer
4. The correction channel

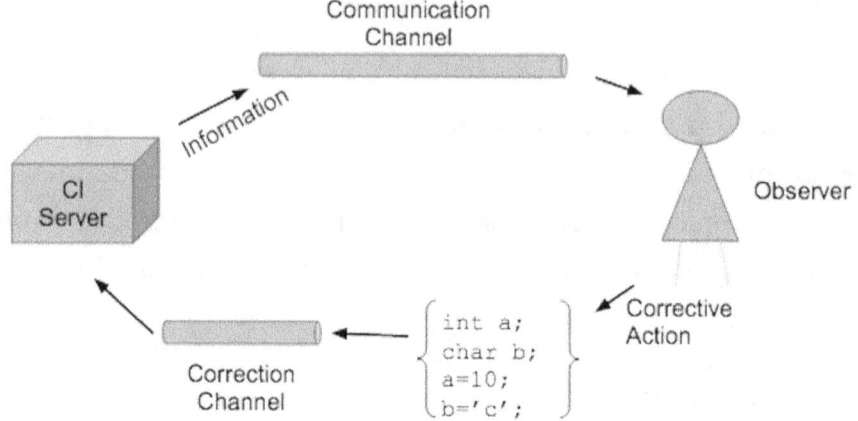

This system assumes that a build has already failed and needs corrective action. The interaction is explained as follows.

1. The build server has the information that the build has failed.
2. The build server passes the information through the communication channel so that it reaches the observer, who is a potential build fixer (not a build breaker because hey, no one is a build breaker!).
3. The observer does a corrective action like changing a line of code, rebooting the database server or anything else that is expected to fix the build.
4. The observer pushes the correction to the build server using a correction channel. This could be a SSH channel for doing a git push, or it could even simply be the build server's user interface to re-trigger the build after rebooting the database server.

## Classification of Subsystems

If we were to classify the system into two subsystems with defined boundaries, we would do it as follows.

1. Communication Subsystem
2. Correction Subsystem

**Communication Subsystem:** This subsystem involves the build server, communication channel and the observer, in respective order. The single goal of this subsystem is to communicate the build failure effectively to the observer.

**Correction Subsystem:** This subsystem involves the observer, correction channel and the build server, in respective order. The single goal of this subsystem is to correct the build failure and push the correction to the build server.

These subsystems are encircled and labelled appropriately in the following diagram.

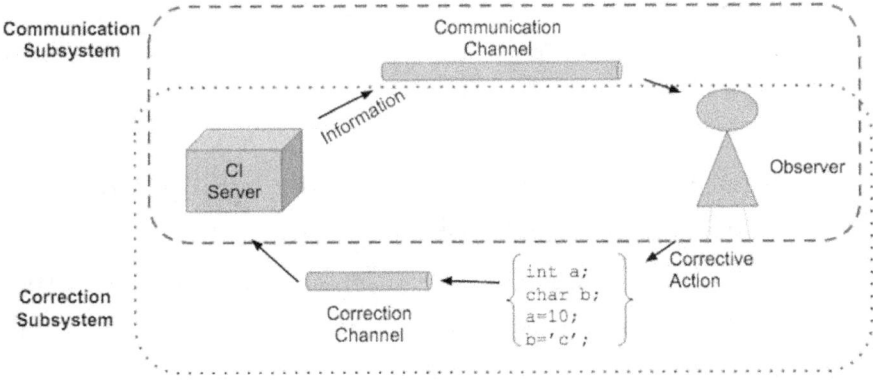

## Possible Reasons for Late Fixes

Why would not a developer (observer) fix a failing build immediately? Why would you have to wait in office at 5PM, clenching your teeth because no one has yet fixed the build that failed at 11AM?

It is very easy for people to assume malice of another person – the assumption that developers are escaping from their responsibility of fixing builds. In other words, it is very easy for people to pin-point the 'correction subsystem' above as an area that needs tuning or improvement. I would not go to the extent of saying that this argument is false. However, I would suggest you to quickly ask yourself the following questions when you are facing a no-fix or delayed-fix situation.

1. Do the developers even know that a build is failing?
2. What is the effort that the developers need to put in order to know about a build failure?

I completely expect a lot of readers to respond to the second question with "Open the email inbox". To those folks, I have another set of questions but now in the 'What if' format.

1. What if the developers are not fixing builds because they do not know that a build is failing?
2. What if developers don't read Jenkins emails because they get frequently spammed by Jenkins with non-actionable emails?
3. What if developers have configured their email boxes to put the Jenkins mails into a separate folder named

'Jenkins' or even worse, Trash, so that they will not show up in their inbox?

4. What if the developers do not want to open their email box while they are busy coding in their IDE?

In short, what if it is just a communication gap that results in suboptimal response times from observers? Will you still ask for tuning the 'correction subsystem'?

Interestingly, the response time of the observer depends on how effective the 'communication subsystem' is. If the communication subsystem is suboptimal, like the email example above, it is only justified to have a suboptimal correction subsystem too.

The rest of this chapter will take you through different ways of configuring the communication subsystem and how that would result in better response times from the correction subsystem.

**Information Radiators – An important concept**

In order to ease the flow of communication, agile teams use Information Radiators. An information radiator is anything that gives out information. When I say 'gives out', it means that the information is pushed to the observer rather than the observer pulling the information from the system with some conscious effort.

So, are all information radiators electronic devices? An information radiator need not necessarily be an electronic device. It can just be some non-electronic media that shows an information.

In order to better understand the information radiators, let me take you through a list of items. Oh, I almost forgot to tell you! You have a task to do. You need to pick out only the items that are NOT information radiators from the following list of six items.

Based on your understanding, which of these are NOT information radiators? (Choose more than one)

1. A clock tower
2. Traffic lights
3. A public announcement system with speakers
4. LPG or Cooking gas pipes
5. Mobile phone that shows the time
6. JIRA Kanban board of a team

Have you chosen your options? Now, let us go through each of these items and check if it is an information radiator or not.

1. **A clock tower:** A clock tower shows some information: the current time. It shows the current time to anyone who is in the range of certain distance, facing it. It does not discriminate between the people. However, anyone who wants to know the time might need to position themselves to be able to look at it. Hence, even if not ideal one, a clock tower is an information radiator.

2. **Traffic lights:** The traffic lights give you a critical information: whether it is safe to go ahead or not. They show the information to any casual onlooker, not just to the people who have subscribed to the traffic light service (There is a reason why there is no service like that.) Hence, this is also an information radiator.

3. **A Public announcement system with speakers:** The PA system transmits the information from the person speaking into the microphone. Anyone who is in the audible range of the speakers can hear that information, without any discrimination. One significant difference from the earlier examples is that this involves the observer using his sense of hearing rather than vision. This is also an information radiator.

4. **LPG or cooking gas pipes:** These things give you signs of danger – when there is a gas leakage. The cooking gas has a distinct smell that can be felt by anyone who is in the vicinity of the leakage. This does not distinguish between the owner of the house and a guest. Here, the observer uses the sense of smell to get the information. This is also an information radiator.

5. **Mobile phone that shows time:** This gives the same information that the clock tower gives. However, there are two important differences – who can see it and how to see it. Since mobile phone is access controlled, only the owner of the mobile phone can see the time. Also, the information is not available all the time – the owner has to pull the information by unlocking the phone. Hence, it is NOT an information radiator.

6. **JIRA Kanban board of a team:** This gives information about the team's current state of work. However, not anyone can see it from JIRA. Only the ones who navigate to the URL of the board can see it, that too, if they have access rights. There is also no sense of 'range' for this system. The information is strictly pull based. Hence, it is NOT an information radiator. However, you can make an

information radiator out of it by displaying it in a monitor in the team room for all to see.

## Information Refrigerators and their use

If the items (5) and (6) are not information radiators, what are they? Don't they also communicate information to people?

If you look at a refrigerator, it does not hand over the food to you. It gives you the food when you open the door. Similarly, the items (5) and (6) above, provide you information only when you open them, typically using your access credentials. So, they are 'Information Refrigerators'.

Information Refrigerators are not useless either. They are excellent for audits. They are excellent for retrospectives. But they are not suitable for real-time communication about events happening 'right now' that need faster responses.

## The suboptimal communication channel

Now, let's get back to our original question of why people may not be fixing builds when email is their only way to know about failures. Do you know the answer now? Yes, email inbox is not an information radiator. It is an information refrigerator that can be used to 'prove' to someone that you had a conversation.

In other words, we were using a suboptimal communication channel (Remember the systemic view?) in the Communication subsystem. One thing that we can try is to add an information radiator to the communication subsystem. This way, our diagram of the CI system will look like the one below.

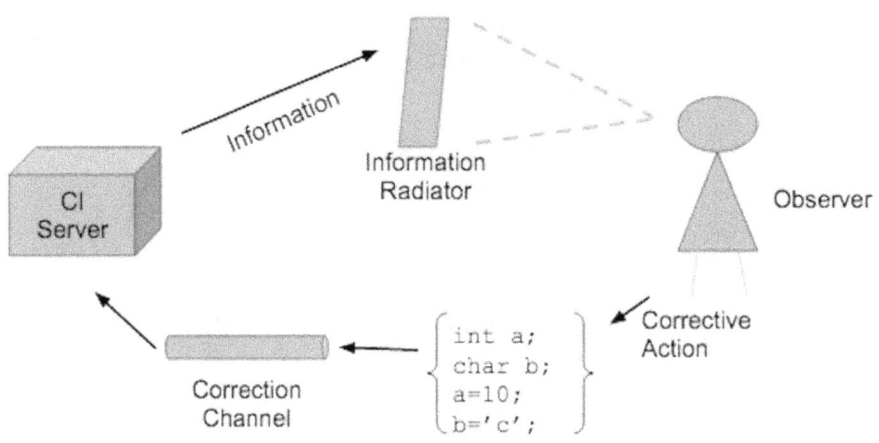

## Sensible Information Radiators

Well, now we have an information radiator in our diagram. What does that translate to in the actual work area? In reality, it translates to many possibilities. What makes an information radiator sensible? For the purpose of continuous integration, a sensible information radiator is something that results in quick response times.

The rest of this chapter will aim to explain the different varieties of information radiators that you can try and how to make them sensible by paying attention to minute aspects of their design, so that the observers respond at the quickest possible time.

## Classification of Information Radiators

Since we are talking about improving the human response times, it makes sense to classify the information radiators based on human cognition. If I classify the information radiators that

I have seen being used in CI environments based on the sense humans use to sense the information, I would create the following categories.

1. Visual
2. Auditory

I am yet to come across a practical information radiator using the sense of smell or touch.

Let us go through practical aspects of each of these categories and ways to make them sensible too.

**Visual Information Radiators**

These are the ones that rely on the observer's sense of sight to push the information about a failing build. Classic example of a visual information radiator is a build monitor.

If you are using Jenkins, you can get a customized build monitor by installing the Build Monitor plugin. This plugin gives you the status of chosen builds in a URL. The build monitor URL allows anyone with access to the Jenkins server to view the status of chosen builds from their browser. But, is this an information radiator? Not until you display the browser screen in the team area for all to see, may be using a monitor or projector.

By displaying the build statuses for all developers to see using a screen in the team's work area, you convert the information refrigerator called Jenkins to an information radiator.

## Cost of Build Monitors

Build monitors need not be costly. You can build an awesome build monitor at a cheaper cost if you use a credit-card sized computer called Raspberry Pi, which is dead cheap, instead of a laptop. You will also find a dedicated section in the appendix towards the end of this book explaining the steps of how to install and configure a build monitor using Raspberry Pi.

Assuming that you have a build monitor available as a visual information radiator, what are the aspects that you can tune to optimize the response times? Let us take a look at two variables and see how the response times can change when we tune them.

Here are the two variables:

1. Position
2. Color Scheme

## Position of Monitor: Does it matter?

Does it matter where you position the build monitor? Can response times of the developers change based on where you position the monitor?

Before we delve into further details, I would like you to take a look at the following team room arrangement and find out the error in it.

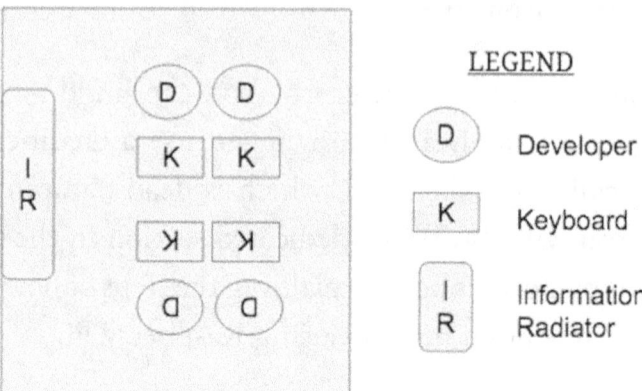

Take a look at the position of the information radiator – the build monitor. Can the developers look at the monitor without turning their heads, during their normal course of work? No, they can't. Even when we have an information radiator here, it is not sensible. It requires a 'polling' behavior from developers to know the status of builds.

So what? What is the big deal if the developers need to turn their heads? This is a big deal because if a build fails, the developers will not know immediately but only at the moment they choose to turn their head. This also keeps them distracted from their work. This is basically an avoidable conscious context switch.

**Is there a better option?**

Yes. There is a better option. The better option involves positioning the information radiator in a place where the developers can just lean back on their chair to know the build status. It is even better if developers' peripheral vision covers the build monitor while they concentrate on their own work in their laptops. Here is an illustration of how we can change the setup to impart these advantages.

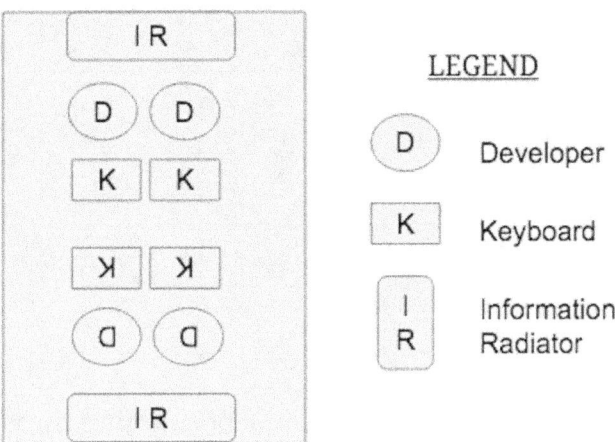

## Isn't this costlier?

Hey, now we have two build monitors instead of one. Isn't it costlier? Yes, it is costly but also effective. The money that you lose by suboptimal response times on an ongoing basis will be more than the one time investment on an additional build monitor.

There is good news if you add more teams to your organization. This design scales well. You will still need two monitors only if you happen to extend the square layout, up to some number of teams. The following diagram shows how two monitors (may be bigger ones) are enough for a bigger workspace with more teams.

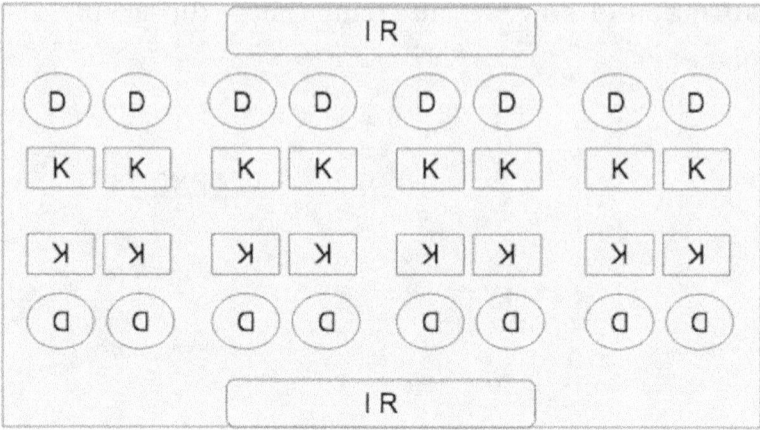

What did we learn from this? Sensibly positioning the visual build monitor will result in faster response times from the developers.

## Color Scheme – An important factor

A build monitor effectively communicates the state of integration builds. It communicates whether a build has passed, failed or in progress. If you see the earlier jargon that we used, we called the failed builds as 'red builds' and the passed ones as 'green builds'.

We borrow the color scheme of build monitors from the traffic light color scheme. The train signaling system, which was a precursor to our traffic light system arrived at the current red-yellow-green color scheme as the third version, after causing much loss of lives and frustration in the initial two versions. The evolution of this scheme is in itself an interesting story for you to lookup as a homework (or in the list of references at the end of this chapter).

Scientifically speaking, red light has a longer wavelength. It can be seen from longer distances. The red color also acts as an evolutionary motivator for human beings to do something aggressive to save oneself. An article in the Scientific American titled "How the Color Red Influences our Behavior" written by Susana Martinez – Conde talks about how the drivers blocked by red colored vehicles acted more aggressively than those who were blocked by the vehicles of other colors. This confirms that using red is indeed a right choice to indicate a failed build, which translates to "Act now! Fast!" to promote the fast build fixes.

**Case of Colorblindness**

The statistics say that one in eight men are colorblind and one in two hundred women are colorblind. This statistic represents the demography across the world. However, local statistics may vary. Doesn't it sound foolish to not consider color blindness into equation when we choose colors for our build monitors?

Colorblindness does not mean that the patients view the world in black and white. It means that they have difficulty differentiating some colors from others. In the most common form of colorblindness, the patients are not able to differentiate between red and green. Yes, *Red. And. Green.* Take a moment to digest this. What does this mean to the notion of sensible design of the build monitor?

You can try simulating the vision of a color blind person by using the Color blindness simulator available for public use at http://www.color-blindness.com/coblis-color-blindness-simulator/

## Handling Colorblindness

If a color blind person (remember 1 in 8 men?) cannot differentiate between a passing build and a failing build easily, the entire purpose of the information radiator is defeated.

Jenkins, though not in the build monitor, uses blue in its main screen instead of green to denote successful builds, perhaps to cater to the colorblind users.

On November 20 2013, a user who identified as Ken logged an issue with title 'Colorblind Palette?' in the Github repository of Jenkins build monitor which was assigned a number #30 [https://github.com/jan-molak/jenkins-build-monitor-plugin/issues/30].

I personally find the conversation thread (open for public) in this issue as the best one I have seen from software developers about colorblindness. After many prototypes, the build monitor now has a checkbox labelled 'Colorblind Mode' in the build monitor configuration.

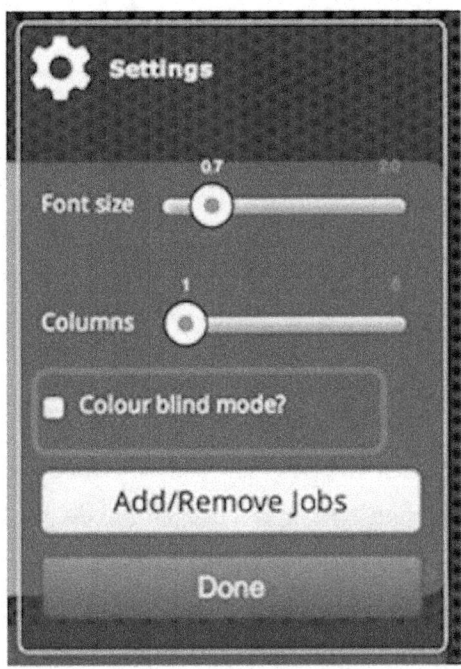

On choosing the 'Color blind Mode', the build monitor then distinguishes failed builds using stripes, which is easy for a colorblind person to notice. One can find a screenshot of the relevant portion of the release notes of Build Monitor Plugin at this web URL – https://github.com/jan-molak/ jenkins-build-monitor-plugin/releases/tag/v1.6%2Bbuild.132

**Auditory Information Radiators**

These are the class of information radiators that convey the news of build failure by means of sound. There is a plugin available for Jenkins which let you upload your own sound files and play them when a build fails either using the browser or using the server's sound card.

If you want to try an auditory information radiator, you might want to concentrate on the following factors.

1. Distinctness from surroundings
2. Uniqueness of the sound
3. Position of the sound emitter

## Distinctness from surroundings

Do you own a mobile phone? You must have configured a ringtone for the mobile phone. Have you ever wondered why people always choose musical ringtones? Why don't we record our own voice, speaking some text and use that as a ringtone?

We use musical ringtones because we want the phone to catch our attention when it rings. If your ringtone is same as the sounds that come from the surroundings, it would not catch your attention. Even when we have dialogues as ring tones, we make sure that there is at least something in it that is different from the surroundings.

Similarly, in your auditory information radiator, you use a sound that is typically not heard otherwise in the development floor. The distinction from the surroundings could even be in the decibel levels. If the build communication is louder than the regular ambient noise, people will recognize the build failures without much conscious effort.

## Uniqueness of the sound

Have you ever noticed that announcements at the railway stations mostly start with a chime or some other music of very short length? Why is that so?

The chime before the announcement prepares the audience that an announcement is about to come. The chime makes the announcer's messages unique and different from other sounds like advertisements that may be playing in the speakers.

If you can include a signature chime for your auditory information radiators just before the announcement of build failures, this serves an important purpose of preparing the audience to listen to the announcement, particularly, if your audio announcement is going to provide additional details about the failure like, say, code coverage.

## Position of the sound emitter

Position is a very obvious thing. It is better to position the sound emitter to cover maximum range so that the sound waves are interrupted by minimum possible number of objects. Typically, the sound emitters attached to the ceilings or columns are the best bet.

If you position the sound emitter a little higher than everyone, this also gives them the intuitive message of considering the build failure announcements as a 'higher voice' that one needs to pay attention to.

## 'Auditory' forwards to 'Visual'

Let us assume that an announcement is made by the auditory information radiator that a build is failing. Developers would want to look at logs in order to find out the corrective action. This means that there should be some visual means of digging into the failure when there is an auditory announcement, without which the response times will suffer.

Auditory information radiator always forwards the attention to a visual information radiator or sometimes a visual information refrigerator like logs.

## Level of Information

Be it a visual or auditory information radiator, what level of information can you communicate? It is best to communicate summary information and not the detailed logs, which are meant for the information refrigerator.

Information radiator just catches the developer's attention. Its job is not to give all information necessary to fix the failing build.

Based on circumstances, you might want to include one or many of the following facts in an information radiator.

1. **Pass/Fail:** This is the bare minimum information provided by an information radiator for continuous integration. I have not come across any information radiator that does not convey this information.
2. **Recent committers:** If people have internalized the Blameless CI culture, then there is additional advantage

in displaying the recent committers list. These are the people who will have the best information or who can get the best information about how to fix a failed build. People can reach out to them for problem resolutions, without attributing blame. If you have just started the Blameless CI practice, I would suggest you not to include committers list.

3. **Qualifiers:** In case you qualify a build as passed or failed based on some other factors like code coverage, it is useful to include these qualifiers (or the failing qualifier) in the information radiator itself. This saves the developer's time which would otherwise be wasted in digging out why the build failed.

4. **Claim information:** Claims plugin in Jenkins helps anyone to claim a build failure, not just the latest committer. This way multiple people will not work on a build failure at the same time. In such cases, it is useful to include the claimer's name in the information radiator.

Remember! Your information radiator just gives summary information. Don't put any detailed information in it because people might consider it to be heavy to look at or listen to.

### What if developers work from their homes?

Work from home as a benefit to attract developers is growing in the startups and some enterprises as well. If your organization lets the developers work from home, then what can you do to the communication subsystem to influence the correction subsystem? Do you even control the communication subsystem for a remote employee?

If you are developing a web application, like many organizations do, your developer alternates between his IDE and the browser multiple times in a day. What if you can put the build failure information on the IDE and browser screens without obstructing the code or the page being browsed?

IntelliJ has a plugin called 'Jenkins Control Plugin' that shows a small, colored icon at the bottom corner of your screen denoting the success or failure of your chosen jobs.

Chrome has a plugin called Monitor Me Jenkins that gives you a very similar functionality in Chrome browser by showing you an icon denoting build failures.

**Keep improving continuously**

We saw some ways of communicating the build failures effectively aiming at faster response times from developers. Are these the only ways? Is this an exhaustive list? By no means can this be an exhaustive list. You can treat these tips as starter tricks and create news ones by trying them and adapting as per the feedback. Remember, we need to pay attention to feedback and change one parameter or the other when the response times are not optimal. This is a journey that is continuous.

**Summary**

Communication about the build failures needs to be efficient for the developers to fix them fast. Sensible design of visual and auditory information radiators can help you in this communication. 1 in 8 men and 1 in 200 women are colorblind and don't leave them out while designing your information radiators.

**Further Readings and References**

1. Systems Thinking – https://en.wikipedia.org/wiki/Systems_thinking
2. Information Radiator – http://alistair.cockburn.us/Information+radiator
3. Raspberry Pi – https://www.raspberrypi.org/
4. How the color red influences our behavior – https://www.scientificamerican.com/article/how-the-color-red-influences-our-behavior/
5. The origin of Green, Yellow and Red Color Scheme for Traffic Lights – http://www.todayifoundout.com/index.php/2012/03/the-origin-of-the-green-yellow-and-red-color-scheme-for-traffic-lights/
6. Statistics on Colorblindness - http://www.colourblindawareness.org/colour-blindness/

# Chapter – 8

# METRICS AND COURSE CORRECTION

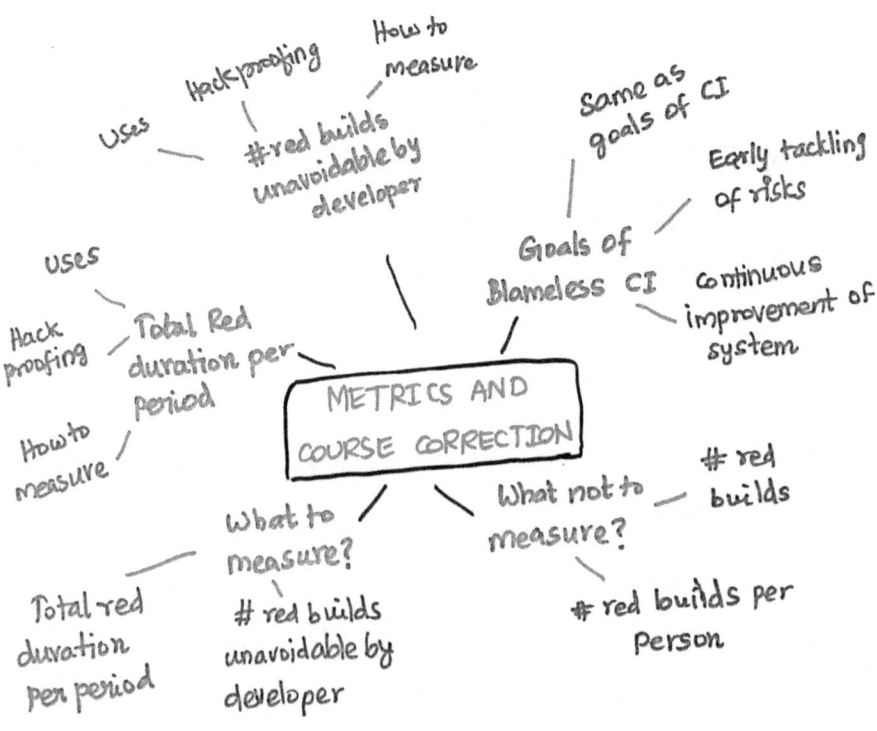

# Where are the mirrors?

Let us say that your organization has started practicing Blameless CI. How do you know if that is causing the intended positive effects to the business? How do you prove to your management that Blameless CI has helped creating business benefits? How do you objectively know if there is actually a culture of fixing builds immediately?

You can tune the system and make course-corrections only when you see the system objectively. This is where some simple measurements can show the mirror to your organization in validating the culture change and the business benefits.

In some cases, few measurements can offer constructive input to your organization in order to prevent future problems.

This chapter is aimed at explaining the things that we need to measure in the context of Blameless CI, how to measure them and more importantly, what the benefits of those measurements are.

## Goals of Blameless CI

What is the motivation for us to practice Blameless CI? From a shallow view, it might look like the motivation is developer delight or engagement. However, if you look deeper, Blameless CI has no bigger motivation than reaping the intended original benefits of Continuous Integration, by adding a behavioral connotation to the same.

In other words, Blameless CI is a practical flavor of CI, different from the one where you could be doing CI using blame games also but with suboptimal results.

While there is an opinion that we can address risks early by preventing integration errors with insistence on communicating more in person earlier, Blameless CI just goes ahead and recognizes that it is not always feasible to communicating about dependencies before code commit. With that notion, Blameless CI tells you it is okay for build failures to happen as long as they are fixed as soon as possible, unblocking everyone.

Blameless CI also helps people to continuously discover systemic loopholes and address them rather than pointing all fingers to a developer's carelessness, without any continuous feedback and optimization of the system as a whole.

In my opinion, the goals of Blameless CI are exactly the same as that of CI namely

1. Tackling of integration risks just in time
2. Continuous improvement of the system as a whole

**What not to measure?**

Before talking about what to measure, I feel the need to talk about what not to measure as this might not be very intuitive. There is a high chance that you might end up measuring a wrong thing in combination with a right thing, ending up in creating a cultural conflict.

In my view, it is advisable not to measure the following things as they add no value or bring negative value to the system.

1. **Number of red builds per period:** It is a common urge to measure the number of red builds in a given period. The common motivation for measuring this is to reduce this number over a period of time by doing communication and due diligence before committing the code. This comes from a basic belief that red builds are bad.

   When you do Blameless CI, you do not believe that red builds are bad. Red builds are good. The red builds drive you early to the action of preventing an integration issue or to challenge your system. If you measure the number of red builds in a period and aim to reduce the number, aren't you shutting your own feedback mechanism? In addition to this, it sends a signal to the engineers that red builds are bad, which will result in non-emergence of blamelessness.

   So, tracking the number of red builds diverts the momentum of Blameless CI to a philosophy opposite to that of Blameless CI.

2. **Number of red builds by a person:** You can easily spot the blunder here. This is the kind of thing that this entire book warns you against. The moment you start attributing the red build to a person, you end up derailing the whole purpose of CI by creating a defensive culture ripe with low commit frequency, bulky commits and less willingness to take risk.

If some organization measures the number of red builds per person while claiming to be blameless, they are just outrightly lying on your face.

If we should not measure the number of red builds with or without attributing it to a person, what else to measure?

In the spirit of keeping the number of measurements as low as possible, I think we can do with just two measurements to validate whether an organization is going in the right direction with their Blameless CI.

1. Total 'Red' duration per period
2. Number of red builds unavoidable by developer per period

## Metric 1: Total 'Red' duration per period

This is nothing but the total duration for which the build was red per period.

Just now I told that we should not measure the number of red builds. Aren't we indirectly measuring the same thing in this metric? No! Instead of measuring the 'number', we are measuring the 'duration'.

How does it matter whether we measure the number of red builds or their duration? Remember that our focus is on de-risking the integration, which happens by 'fixing' the red builds. When you measure the duration of red builds, you are measuring how fast the developers fix the build. Fixing is the behavior that you want to encourage, not avoiding the failures.

Your aim as an organization is to keep this number as low as possible. You can start sensitizing this number as the number of hours a potential tester or developer is blocked due to unavailability of testable/working code. Hence, the beauty of this metric emerges when you start presenting the trend data of this metric over a period of time to the teams and the leadership.

In short, this metric verifies if there is a thriving build-fixer culture and forces people to focus on fixing builds as soon as possible.

**Making the metric hack-proof**

When there is a metric, it is essential to make the metric hack-proof to attain the intended results. The most common example is code coverage as a metric. Developers can easily hack your code coverage metric by writing useless test cases just to cover your code. I have even seen test methods with name coverUncoveredCode().

Good thing about this metric, the total build red duration per period, is that it comes with a way to detect people hacking the metric. With the intention that we need to keep this number as low as possible, people might get into near zero downtimes.

If we have near zero build red duration consistently over long periods, one or more of the following things are true.

1.  Integration builds are not failing. Do we even have a need of integration?

2. Is there a blame culture still prevailing, leading to people not committing code often? Cross-validate this with commit frequency and size of each commit.
3. Are the teams not writing integration tests due to pressure or local optimization (this is a politically correct term used to refer to office politics) inside divisions?
4. Are we actually building disjoint products and trying to think of them as a single product?
5. Are we avoiding building integrations for the fear of something? Remember! Integrations among different features are the ones that make a software viable in the long term.

If you have information radiators installed in your workplace, you can validate whether your own observation of reality is reflected in these metrics.

**How to measure this?**

This is a measure that you can get directly from Jenkins by collating the build results for a period. With little effort, you could write a small script that calculates the build red time for any duration by accessing Jenkins REST API. You could even ask for volunteer developers to create this script as an engineering backlog item.

If you are okay with an approximate measurement, even without a number, you can directly use the Jenkins Global Build Stats Plugin. This plugin gives you a graph of build states over time. It denotes the passed builds as 'blue'. Hence, we need to concentrate on the area between start of 'red' area and the following start of 'blue' area (including white area) in the graph and try to reduce it over a period of time.

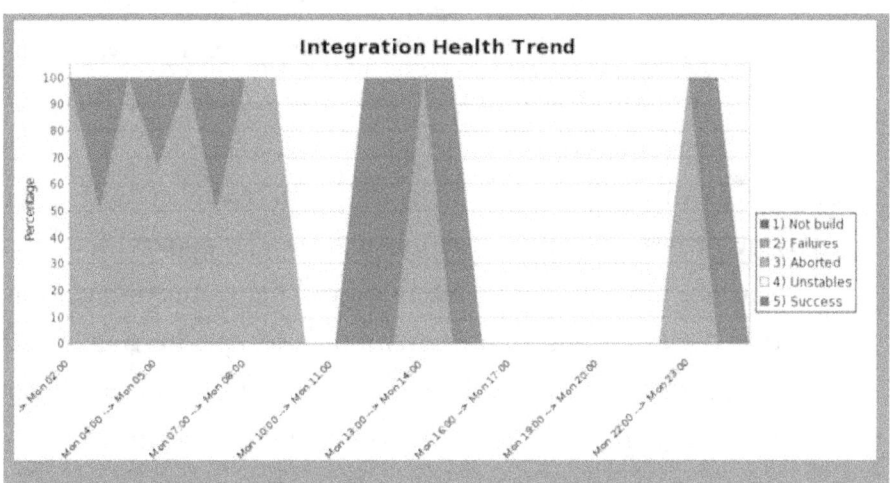

**Metric 2: Number of red builds unavoidable by developer per period**

Remember the way we classified the red builds? Earlier, we classified the build failures into two categories namely

1.  Avoidable by developer
2.  Unavoidable by developer

One thing that we can measure is the number of build failures that happened to be unavoidable by developer. A failing build is classified as (1) or (2) after someone takes a look at the reason of failure. Who does that classification is not a big concern as long as they are professional.

What will we get by measuring how many unavoidable red builds we produced in a period?

1. This is the actual Return on Investment (ROI) of your continuous integration practice. This is the number of things that could have gone wrong at the end of release when integration is done late in the cycle. This validates your case for continuous integration. Show this number to your leadership.

2. If this number is too high or it increases steeply over time, it is an input to the organization in general and the product architecture in particular. This trend proves that your code has high entanglement and can be refactored to accommodate interfaces to suit dependencies. Your architect will be happy with you for bringing this metric's trend to him. He can use this data to get time for refactoring the code.

3. A very high value can also mean that your organization's teams are not organized optimally and this is sign to think of team reorganization. This is a feedback to the management, which is based on objective data.

4. When you trend this data, you can even verify if the re-organization of teams or the refactoring helped by checking if the number went down after the said change.

## Making the metric hack-proof

What does this number mean if this value is near zero consistently? Yes, *consistently* and not just for one or two periods. It might mean any of these things.

1. Due to management pressure, people are classifying unavoidable issues also as the ones avoidable by developers. Remember! Unavoidable failures increase the work for the

management and architecture and not necessarily for the teams. You can get this insight by correlating this with the total build red time (the earlier discussed metric) for same period. If the build red time is not zero but this metric is near zero, assumption is that people are doing wrong classification.

2. This might also mean that you do not need integration per se. Your product does not need integration due to the way people and code is organized. You can validate if it is true by correlating this with the total build red duration for the same period. If this is also zero, you are most likely in the situation where we do not need integration. In some cases, it could just be that people are not writing enough integration tests due to lack of focus and time.

This way, you can validate if anyone is trying to hack this metric by taking shortcuts by correlating it with some other fact.

**How to measure this?**

Before measuring this, you need to get a way to categorize the build failures. You can use Build Failure Analyzer plugin where anyone can create categories as reasons for failure and change them after the build broke. This is not the main use case of this plugin but you can use the plugin this way to measure this metric.

Now that classification is done, you can develop a script that fetches the build data using REST API of Jenkins and calculates the number for current period.

With just these two metrics, you can get valuable feedback for course corrections both at the team level and the organization level while you are in the continuous journey of Blameless CI.

## Summary

Goals of Blameless CI are that of CI itself. Blameless CI is a practical variation of CI. Total build red time per duration is used to identify whether there is a culture of build fixing. Number of failures unavoidable by developer per duration is used to identify whether CI is needed in the first place and also to demonstrate the ROI of the CI practice.

## Further Readings and References

1. Martin Fowler's article on appropriate use of metrics and what not to measure – https://martinfowler.com/articles/useOfMetrics.html
2. Global Build Stats Plugin – https://wiki.jenkins-ci.org/display/JENKINS/Global+Build+Stats+Plugin
3. 7 Metrics to track when implementing Continuous Delivery – http://www.datical.com/blog/7-metrics-to-track-when-implementing-continuous-delivery/
4. Continuous Integration Build Metrics – http://www.thinkinginagile.com/2011/07/continuous-integration-build-metrics_16.html

## Parting remarks

This chapter marks the end of this book. This book is just a collation of my limited understanding of the subject based on my experience. I would go to the extent of saying that this might just be a beginning of my ideas by mingling two different domains – Technology and Social Psychology.

I am sure you will find ways to improvise or customize the theory and practices in this book to make it better. Share your learnings with your professional network and to the public. Tweet your thoughts and learnings (#BlamelessCI may be?), discuss your learnings with your friends because discussions, coupled with personal stories, take the word out to the masses.

In the spirit of continuous improvement, I would like to keep the feedback channel open. Please send your feedback to me. I will feel happy and honored to hear from you.

Oh, yes! There is an appendix too, which consists of articles that you can read along with this book's core content.

# APPENDIX-A

# 21 INSTANCES WHERE A WRONG PERSON GETS BLAMED FOR A BUILD FAILURE

In most of the teams that I have worked with, the build breaker is frowned upon, sometimes treated like a criminal. Until recently, even the Jenkins build monitor called the person who committed just before a failing build as a 'culprit'.

Though there is no denying that developer's inattention to detail causes some build failures but developer's negligence is not the only cause of a build failure.

This drove me into writing a non-exhaustive list of ways, based on my experience, in which the wrong person gets blamed for a build failure just because he/she happened to be the last one to push a code change.

## Assumptions

### a. People Involved

Here are the names that I would like to give to the people involved in illustrations.

I – Myself, the sorry soul who gets blamed for the build failure
X – Previous Committer, who pushed a commit after my checkout

### b. Steps followed by developer to push new changes (assuming git)

1. I checkout the code
2. I make changes
3. I run tests
4. I commit
5. I push my commit to upstream (Rely on IntelliJ or IDE to rebase my commit on top of any commit that came in after my checkout – Essentially do a git pull and then a git push)

REMEMBER! It is very important and urgent for me to push as soon as the rebase happens or else someone will commit in the meantime and I have to rebase again. If you understand this fact clearly, there is no need to struggle to understand any items in the following list.

The situation gets even more complicated if I have a pre-commit code review system like Gerrit, where I do not have control on how long it takes for a reviewer to give an approval.

Now that you know what the context is, here are the 21 instances. Oh wait! Remember! This is not an exhaustive list.

## 21 Instances Where ...

### Compilation Errors

1. I add a variable 'myVar' for some purpose. X also added a variable 'myVar' for a different purpose. This applies to names of class, method or any other identifier.
2. X deleted property/method of a class that my code uses.
3. X added an additional parameter to a method that my new code calls. This applies to removing/reshuffling parameters too.
4. X upgraded a library that the code is dependent on. My code calls a method that is unsupported in new version but supported in old version. This applies to downgrading versions too.

### Test Failures

1. I added a new test. X pushed code that does not pass my new test's expectations
2. I write a test, taking some pre-conditions for granted, because that is always the case so far. X pushed a test, which while running, will mess with the expected pre-conditions of my new test. For example, test written by X might change a global variable to a value that will cause my new test to fail. This is not an issue in case of tech stacks which provide clean slate for each test.

3. I write a UI test that opens a dialog and checks its contents. X pushed a test that opens a modal dialog but forgot to close the modal dialog at the end of the test. This would prevent my test from even opening up my expected dialog.

4. I wrote an integration test that traverses the following classes: A-> B-> C. I wrote mocks for C. Meanwhile, X committed a test that mocks class B and let his test share my test's environment. This is not an issue in case of tech stacks which provide clean slate for each test.

## Incorrect CI Configuration

1. Recent breaking change in CI Job configuration. CI job generates an artifact/file based on the source files that I have in repository. I deleted a source file in the last commit. Expectation is that the generated file is also not present during building.

   But, CI job's workspace still contains the last generated copy of the artifact, even if the source file is deleted from the repository. For example, I delete a class but my CI job has the .class file that was generated from the earlier build. Same is the case with typescript compilation from .ts files to .js files. This is usually a case where 'not clean' or incremental builds are performed (which is actually a good thing in some cases).

2. Till date, CI has been showing 'green' all time because of misconfiguration. Just before I pushed my commit, the CI job was corrected to show the actual results.

3. Another job is configured to use/modify the directory that my job also uses. Name-spacing/isolation issues.

## Build Environment Issues

1. The test database ran out of space just before my commit. This does not apply to cases where a database is created fresh with every build.
2. The CI server ran out of disk space just before my commit.
3. The database/external service used by the CI job to run the build/tests was upgraded just before my commit.
4. Changes were being done on Access Control Lists of environments during the time my commit was being validated. AWS has forcefully terminated the EC2 instances used by CI job for some reason just when my commit was being validated. (For example, doubtful malicious activity, over dues, etc)

## Just Setting up – Starting Troubles

1. The step to run tests in CI job was enabled just before my commit
2. X committed a test configuration file (especially something like karma.conf.js file) that would include my new files and fail when my new files are run inside tests.
3. CI job was using a trial version of a software involved in the build process till date. Now upgraded to paid version which enforces more stringent standards.
4. New rule added to SONAR (or any static code check) just before my commit.

There are ways to avoid a lot of the above wrong failures. But, rarely, many of them fall under the control of an individual developer. These preventive measures either end up with architects or the maintainers of CI environments, who, in fortunate cases, are part of development teams (Devops anyone??) and in most cases not.

Moreover, this list is not even exhaustive. I wrote this list from my context of work. I am sure the readers will come up with scenarios that fit their individual contexts.

## What do we learn from this?

The single piece of learning that I would take away from this list is this:

*Blaming the last committer for a build failure, without looking into the details, is just plainly gross and offensive!*

It is more efficient and cheaper to concentrate on 'fixing' builds rather than concentrating on ostracizing the build breaker. What are your thoughts on this?

# APPENDIX-B

# LOW COST BUILD MONITOR
# USING RASPBERRY PI

**Why should you setup a Build Monitor?**

In the spirit of Blameless CI that I have been advocating in my recent posts, the first and foremost step in empowering people to fix builds is to let them know that the build is failing now. A build monitor helps you to do just that.

**Don't we already have other ways to communicate build failure?**

Jenkins has a beautiful Build monitor plugin that you can access from a URL but you cannot expect all your developers to keep this URL open all the time in their computer.

Email communication for every failed build is a good idea. However, in Blameless CI, since anyone is empowered to fix a build, every possible developer should be notified about a failing build. Hence, build failure mails are more likely to be "filtered" and discarded if you have a huge target audience. Nobody wants a bloated inbox.

If your teams are working from an office, a conspicuously placed build monitor will do the job of communication for you.

## How much will a Build Monitor cost?

Why spend a lot on buying a laptop or a desktop computer when you can get a cool build monitor for one-tenth of the cost of a laptop? Raspberry Pi is a cheap and powerful credit-card sized computer that runs Linux. Though it is better suited for Internet-of-Things (IoT) prototypes, Raspberry Pi can also be used to quickly install a build monitor for your teams, if you are into Continuous Integration (CI).

Here are the costs.

Raspberry Pi Kit – 4000 INR (approx.)
HDMI Monitor – 5000 INR (approx.)
-------------------------------------------------
Total – 9000 INR

## Raspberry Pi – Minimum Configuration for a Build Monitor

Buy a Raspberry Pi with following accessories

- Minimum 3 USB Ports for Keyboard, Mouse and Wifi
- Adapter
- Wifi Adapter/Ethernet Port
- HDMI Port
- SD Card – 8 GB

## SD Card? But, where is the Hard Disk?

Raspberry Pi works with an SD Card as a memory device instead of a hard disk. So, the Operating System and your files will go into the SD Card.

SD Card comes pre-loaded with the Raspbian OS, a Debian variant of Linux. If not, visit this page to know how to flash the OS into the SD Card. You might need a computer with an SD Card reader to do that.

## Configuration – Steps

### 1.   Boot to Desktop with auto-login

If your Raspberry Pi automatically boots to the graphical desktop, you can skip this step. If not, you have to perform this step.

Run the following command after logging in as pi user.

sudo raspi-config

In the menu that appears, choose "Boot Options" > Boot to desktop as pi user

Restart after selecting this option. Now, the Pi should boot to desktop directly without asking for password.

### 2.   Connecting to Wifi

Ease of wifi connectivity has increased with newer releases for Raspbian. Now, with the latest Raspbian you can connect to your Wifi network very much like you do in Ubuntu Linux.

Network icon in the panel lists the available networks on clicking and you can connect to your network by entering key, if any.

If you have older raspbian versions, I would recommend to upgrade to latest version to avoid complications in Wifi connectivity. Any additional help about connecting to Wifi can be found here.

### 3.  Installing necessary software

You will need the following software to run the build monitor.

**xautomation**: This is a package to perform fake keyboard/mouse input. This will be used to simulate the F11 key press to activate full-screen mode in your browser.

**epiphany**: This is a light-weight browser that runs on linux. This comes pre-installed with Raspbian. But, you might need to install explicitly if that is not the case.

Run the following command in a Terminal window to install xautomation and epiphany in a single go.

sudo apt-get install xautomation \

epiphany-browser

### 4.  Startup Script to open browser with Jenkins build monitor URL on login

Use the text editor to create a file startup.sh in /home/pi directory with the following content.

```
epiphany-browser -a \ http://<<my_build_monitor_url>> --profile
\ ~/.config &
sleep 15
xte "key F11" -x:0
```

## Where do I get the URL to fill in the first line?

The first line of the code opens epiphany browser with the given URL. In order to get this URL, you must have a build monitor view setup in your Jenkins server. Copy your build monitor's URL and paste in the first line of the file.

## What's with the ~/.config parameter?

The first time usage will require you to login to Jenkins using keyboard. But, you don't need to re-login in next usage. That is because, in the first line, we are also instructing the epiphany to store the session data in ~/.config directory so that you need not re-login to Jenkins when the raspberry pi reboots again.

## What does xte do in the third line?

Do you remember the xdotools that we installed? We use that in the 3rd line to simulate a F11 key press to change the display of browser to fullscreen.

## Permissions

In order to execute this file as a shell script, you need 'x' permission bit set. Enable execution by running the following command.

chmod +x /home/pi/startup.sh

## Hooking up the startup script to login

In order to instruct the pi to run your startup.sh during login, you need to edit the /home/pi/.config/lxsession/LXDE-pi/autostart file.

Run the following command to open the above file in nano text editor.

nano \

/home/pi/.config/lxsession/LXDE-pi/autostart

Add the following line as the last line in the file and save it.

sh /home/pi/startup.sh

The above command will run the startup.sh file when auto-login happens

## 5.   Scheduled Cron job to restart the device every 15 minutes

There are hundred different things that could go wrong with your build monitor – be it network issues, Jenkins availability issues, etc. If you are required to be near the pi all the time, it is counter-productive. Hence, it makes sense to program the pi to restart at regular intervals.

In my environment, I restart every 15 minutes. But, feel free to choose your interval based on your environment.

In order to schedule the restart, we need to use cron jobs. Type the following command as pi user and press enter.

crontab -e

Choose your favorite text editor if you see a menu. If you do not understand the editor names, choose nano to be safe.

In the file that opens, add the following as the last line and save.

*/15 * * * * sudo reboot

Choose the cron expression based on your decided time interval.

**Pro Tip:** If you are installing multiple build monitors to be placed in different areas of your office, make sure that the restart schedules of the PIs differ from each other. That way, you will not end up in a situation where you see boot screen in all the build monitors. At least one build monitor will be showing build status.

**How to test my work?**

Simple! Just reboot your pi using the menu or using following command.

sudo reboot

*Tada*! You have a low cost build monitor in your team room now.

# APPENDIX-C

# TACKLING THE BLAME GAME DRAGON

**A scene during a retrospective**

"How can you expect me to reduce the 'blocked' time for the testers? No one cares about red builds. Developers take days to fix the builds. They are simply shirking their responsibilities. Behavioral issues need attention and action from authority," a voice complained in the program retrospective meeting. Obviously, a root cause analysis was in progress.

After a heated exchange of words, a decision (irrelevant to us now) was reached. Someone noted the action item. A few minutes passed, and a few more issues were discussed.

The same voice spoke again.

"One good thing that happened in this iteration is that we were able to get the [insert tech talk here] standardized for developers across the program, thanks to the tireless efforts of my team members."

Maybe you've heard something like this before. Do you see the problem? No? Read on!

Let me assume the voice of the 'the voice' above and rephrase what was said.

**Situation 1:** I am not able to influence the developers to fix red builds faster. It is the problem of developers. Escalate to authority.

**Situation 2:** I (or my team) was able to influence developers to do X. It is my (team's) achievement. Reward me.

Now you see the problem, right?

**Understanding the Fundamental Attribution Error**

Take a minute and try to recall a time when you behaved like the voice above. If we are honest with ourselves, we probably can think of at least one instance. Everyone commits this mistake. It even has a formal name in social psychology — a Fundamental Attribution Error. A fundamental attribution error is just another way of saying, "Something went well because I was the reason. Some other thing went badly because the situation was bad." It is a mistake that almost every normal human commits.

This error might just be secretly killing your team culture, without being very visible. The long-term effects of ignoring the fundamental attribution error could be that the team stops talking about uncomfortable topics, i.e., their mistakes, because they know that the person with the loudest voice will never accept any part in the mistake. This can seriously hamper the learning system of the team.

## Digging deeper

Let's dig deep into what can qualify as the reason for something going well or badly. When something goes well or badly, the reason can be one of the following:

1. Person/people
2. Surroundings/situation
3. Timing

Unless one pays great attention, it is easy to decide on the most visible or salient item in the list above as the cause. For example, it would be easy to blame a new member of the team for the team's high bug count because the addition of the new member is more 'salient' or visible in relation to other issues that might have caused a high bug count. The effect of salience also plays a role in the Fundamental Attribution Error.

## Objectively determining the real root cause

If all human beings are vulnerable to the fundamental attribution error, what is the right way to arrive at the real cause of a situation? Luckily, social psychology also provides us with a way to differentiate among these three possible root causes – people, situation or the timing, as listed above. A model that helps us to zero-in on the actual root cause objectively is called the Covariation Model.

During the root cause analysis, to wear an objective hat, keep looking for the following factors:

1. **Consensus:** How many people/teams do it?
2. **Distinctiveness:** Do all situations result in such behavior?
3. **Consistency:** Does this happen repeatedly or was it just this one time?

Let us take an example of the "build red" situation and explain the above terms:

## Do all developers not fix the builds or only some developers? (Consensus)

1. All developers (High consensus)
2. Only some developers (Low consensus)

## Do they not fix builds only in the current notification system, or even when alternate notification systems are installed? (Distinctiveness)

1. Fix builds in alternate notification systems but not the current one (High distinctiveness)
2. Not fixing builds irrespective of notification system (Low distinctiveness)

## Was this behavior seen only in this iteration or was it in the past iterations, too? (Consistency)

1. Past iterations too (High consistency)
2. Only this iteration (Low consistency)

We just have to follow the matrix below to objectively arrive at the real root cause.

| Consensus (Person) | Distinctiveness (Situation) | Consistency (Time) | The "real" root cause |
|---|---|---|---|
| Low (Not all) | Low (Both in current notification system and in other systems) | High (Every time) | Person/People |
| High (All devs) | High (Only in current system) | High (Every time) | Situation (or notification system, in this case) |
| High (All devs) | Low (All systems) | Low (Only this time) | Time (Probably this iteration, the team is under high pressure to do more features) |

Take some time to digest the above table. Yes, it takes a few minutes!

Now, you might ask me, "What do I do if I get values like Low, Low, and Low in all three variables?" There is no answer for such a combination in the above table. In such a case, the simple answer is that we do not know. There is not enough information to arrive at a single root cause. Instead, try attacking two possible causes.

By following the Covariation Model in root cause analysis, particularly during retrospectives, we can unlock some powerful perspectives. Make this a habit, and the team is more likely to openly discuss problems and solve them when empowered by the objectivity that this model offers.

This article was published first in ScrumAlliance member posts at the following address:

http://www.scrumalliance.org/community/articles/2016/august/ taming-the-blame-game-dragon

www.ingramcontent.com/pod-product-compliance
Lightning Source LLC
Chambersburg PA
CBHW051534170526
45165CB00002B/724